DISTILLATION - INNOVATIVE APPLICATIONS AND MODELING

Edited by **Marisa Fernandes Mendes**

Distillation - Innovative Applications and Modeling
http://dx.doi.org/10.5772/62970
Edited by Marisa Fernandes Mendes

Contributors
José Del Carmen Zavala Loría, Asteria Narvaéz García, Alejandro Ruiz Marín, Yunuen Canedo López, Adel Zrelli, Mª Mercedes Del Coro Fernández-Feal, Luis R. Sánchez-Fernández, Blanca Sánchez-Fernández, Rosenberg Javier Romero Domínguez, Vilmar Steffen, Edson Antonio da Silva, Nelio Teixeira Machado, Nermina Spaho, Adriana Téllez-Anguiano, Mario Heras-Cervantes, Juan Anzurez-Marín, Gerardo Marx Chávez-Campos, Jose Antonio Gutierrez Gnecchi, Kuo-Ting Wang, Wei-Hsiang Lai

Published by InTech
Janeza Trdine 9, 51000 Rijeka, Croatia

Publishing Process Manager Diana Olloni
Technical Editor SPi Global
Cover InTech Design team

First published June, 2017
Printed in Croatia
Legal deposit, Croatia: National and University Library in Zagreb

Additional hard copies can be obtained from orders@intechopen.com

Distillation - Innovative Applications and Modeling, Edited by Marisa Fernandes Mendes
p. cm.
Print ISBN 978-953-51-3201-1
Online ISBN 978-953-51-3202-8

For EU product safety concerns:
IN TECH d.o.o., Prolaz Marije Krucifikse Kozulić 3, 51000 Rijeka, Croatia,
info@intechopen.com or visit our website at intechopen.com.

Meet the editor

Professor Marisa Fernandes Mendes, born in 1972 in Rio de Janeiro, Brazil, is a full-time professor at the Federal Rural University of Rio de Janeiro (UFRRJ), Chemical Engineering Department, situated in Seropédica City, Rio de Janeiro State, Brazil. She has over 15 years of research experience in the fields of experimental and modeling, simulation, liquid-vapor equilibrium measurements, thermodynamic modeling, biofuels, and separation processes (supercritical fluid extraction, distillation, adsorption, bioreactors). All of her research has been developed in the Applied Thermodynamics and Biofuels Laboratory in the Chemical Engineering Department, where she has been performing her pedagogic activities since 2005. Professor Mendes graduated in 1996 from the Federal University of Rio de Janeiro (UFRJ), and obtained her D.Sc. in 2002 from the Chemical Engineering Program (COPPE/UFRJ). In 2003, she completed her postdoc at the same university, and two years later she became a full-time professor. Moreover, Professor Mendes completed another postdoc at the University College London, United Kingdom, in 2015, applying mathematical modeling to study the behavior of biological systems. Prof. Mendes' research has resulted in 60 journal publications and over 150 conference publications.

Contents

Preface

The art of distillation dates back to at least the first century A.D. [1, 2]. This state of the art process is the heaven that column designers always dreamt of [3]. It is not uncommon to see in books on distillation, the design of columns to separate hydrocarbons, and the mass production of mathematical models and correlations to represent the separation process. This is due to the fact that distillation today accounts for 90% and 95% of all industrial separations. Approximately 40,000 distillation columns are in operation in the United States alone, requiring around 40% of the total energy consumption in U.S. chemical process industries [4].

According to Górak and Sorensen [5], distillation is the major consumer of energy in process industries globally, and because of that the amount of research dealing with energy efficiency has increased, as well as with the economic evaluation of the process itself. Connected to all of these aspects, the distillation process is complex and requires adequate modeling, a challenge for researchers and scientists around the world.

Although several books are available to study distillation concepts, new applications and mathematical models proposals can only be found in journals concerning chemical engineering problems.

In summary, this book is composed of chapters that discuss different applications of the distillation process, linked to areas such as economic evaluation, studying and demonstrating energy saving, mathematical modeling, and adjusting and correlating experimental data to better predict and design new distillation columns.

Marisa Fernandes Mendes
Chemical Engineering Department,
Federal Rural University of Rio de Janeiro (UFRRJ),
Seropédica City, Rio de Janeiro State,
Brazil

References

[1] Forbes, R. J., Short History of the Art of Distillation, E.J. Brill, Leiden, 1948.
[2] Seader, J. D., Henley, E. J., Separation Process Principles, John Wiley & Sons, 1998.
[3] Kister, H. Z., Distillation Design, McGraw-Hill, 1992.
[4] Wankat, P.C., Separation Process Engineering, third ed., Pearson, 2012.
[5] Górak, A., Sorensen, E., Distillation – Fundamentals and Principles, Elsevier, 2014.

Section 1

Column Process Design and Calculations

Chapter 1

Steady-State Modeling of Equilibrium Distillation

Vilmar Steffen and Edson Antonio da Silva

Additional information is available at the end of the chapter

http://dx.doi.org/10.5772/66833

Abstract

In this chapter, an algorithm for the solution of the mathematical model featuring steady-state multicomponent distillation columns is analyzed and applied in the case study of the separation of hydrocarbon mixture. The development of the model has assumed each stage outlet streams in thermodynamic equilibrium in the phases liquid and vapor. The modeling of liquid was considerate and the non-ideality behavior was described by activity. The non-ideality of gas phase was calculated by Peng-Robinson equation of state. The model consists of a set of nonlinear algebraic equations. The algorithm and numerical procedure to solve a set of equations are presented in a sequential, general and very simple form. A methodology to produce the good initial guess was defined based on rude simplifications of the system. In the study case, the initial estimates generated by the method are very good, being only about 20% far from the simulation results and considering a tolerance of 10^{-10}, the convergence was obtained with 28 iterations.

Keywords: distillation, steady state, phase equilibria, modeling, simulation

1. Introduction

The analysis of a plant, by simulation, within the development of new processes may frequently show beforehand whether it is technically and economically viable. The simulation process in already operating plants may optimize the operational conditions for better quality products, decrease energy consumption and other losses in the process [1].

The design of a multicomponent distillation column by phenomenological models is quite complex due to the large number of parameters and variables involved [2] and also usually, it is required to solve the set of nonlinear equations and differential equations. Mathematical modeling is a powerful and useful tool in the design of this type of equipment, it assists the control and optimization column and therefore, the project and operating costs can be significantly reduced.

The use of distillation as separation method is disseminated by the modern chemical industry. One can find it in almost all industrial chemical processes where liquid separation is required. Common commercial binary distillations are as follows: water/ethylene glycol, benzene/toluene, o-xylene/m-xylene, isopentane/n-pentane, ethylbenzene/styrene, water/acetic acid, ethanol/water, among others [2]. There are a lot of hydrocarbons mixtures that can be cited as commercial multicomponent distillation examples.

With all the foregoing, it is clear the need to provide the theoretical study on a simplified model for evaluating the possible separation processes using distillation columns. The simplest mathematical model for a distillation column is obtained, considering that all stages outlet streams (liquid and vapor) are in thermodynamic equilibrium. What represents, with no much accuracy, what happens in an actual process; however, this study is very important to get an idea of the theoretically best result that can be achieved in the process in question.

Thus, the equations of a distillation column model are obtained from mass and energy balances, mass balances by component and iso-fugacity equations. The equations that represent such model are highly nonlinear, particularly those describing the phase equilibria and energy balances.

The solution of a set of nonlinear equations is quite difficult and generally requires that good initial guesses are provided in the way the method presents convergence [3]. Thus, the solution of the obtained model is divided into several steps, where in each step is calculated a set of model unknowns (mole fraction, temperature, flow rate, etc.), such that it is not necessary the solution of nonlinear equations sets, but only sets of linear equations, evaluation of explicit algebraic expressions and root finding of single variable. Also will be presented a methodology to generate good initial guesses and an example will be studied using the methodology presented.

The simplest methods used for solving the modeling of distillation columns are the graphical ones like McCabe-Thiele [4] and Ponchon-Savarit [5, 6]. The equation obtained by modeling the steady state of equilibrium distillation columns forms a set of highly nonlinear equations (MESH equations, obtained from mass and energy balances, phase equilibrium relations and mole fractions summations) that are normally solved all at the same time by Newton-Raphson and like-one method. Another, much used, type of method is that one which uses "tearing equations," that is, large sparse systems of algebraic equations are split into smaller systems and solved in a sequential form [7–11]; it makes the solution process simpler but, as a consequence, can occur some instability. So, the idea of the method being proposed aims to join the efficiency of Newton-Raphson-like methods with the simplicity of methods based on tearing equations, what is possible only with a very good initial estimates generating methodology.

2. Model assumptions

The following assumptions were made when formulating the model of the distillation process:

- Steady state;

http://dx.doi.org/10.5772/66833

- No reaction occurs in the column;
- The vapor and liquid phases are homogeneous in all stages;
- The vapor and liquid leaving any stage are in phase equilibrium;
- Heat transfer only on condenser and reboiler, unless otherwise specified;
- The model does not include effects due to column internals (e.g., pressure drops and flooding/weeping).

3. Modeling

The modeling of a steady-state distillation column is based mainly on mass and energy balances; in this way, it is needed to understand the equipment layout to obtain such mathematical equations [8, 12, 13]. This model is based on the equations of the column called MESH (material balance equations, phase equilibrium equations, mole fractions summation equations and heat, which means energy balance equations). Aiming to make the model as general as possible, it will be considered that can exist feed stream in any stage and the output streams (bottoms and distillate) can be in liquid phase, vapor phase or both phases [14]. **Figure 1** shows a schematic representation of a distillation column and **Figure 2** shows a schematic representation of input and output streams in a stage.

Where F_j is the flow rate of the feed stream to stage j, Q_j is the heat load from stage j (for convention, it was assumed the heat leaving the stage), L_j is the liquid flow rate outputting stage j and inputting stage $j+1$, V_j is the vapor flow rate outputting stage j and inputting stage j–1, U_j is the liquid side flow rate outputting stage j and W_j is the vapor side flow rate outputting stage j. The side streams U and W are used to represent the output streams like distillate and bottoms, that is, D and B. These streams can occur as liquid (L_B and L_D) or vapor (V_B and V_D), in which $D = L_D + V_D$ and $B = L_B + V_B$. We are going to consider n stages, with the stage numbering starting on condenser ($j = 1$) and continuing until reboiler is reached ($j = n$). The generalization from **Figure 2** to represent all stage is shown in **Figure 3**.

Considering the schematic representation of **Figure 3**, it is possible to consider side stream in any stage, not only in the first but also in the last stages. A little care is needed in utilizing **Figure 3** as a base for obtaining the mass and energy balances, that is, how there is no stage 0, so the streams U_0 and L_0 that would output and stream V_1 that would input this stage do not exist; similarly, how there is no stage $n+1$, so streams V_{n+1} and W_{n+1} that would output and stream L_n that would input this stage do not exist. So, it is needed to fix $W_{n+1} = V_{n+1} = L_n = U_0 = L_0 = V_1 = 0$.

So, one can obtain the mass balance in stage j as

$$F_j + V_{j+1} + L_{j-1} - V_j - W_j - L_j - U_j = 0. \tag{1}$$

The sum of mass balance in all stage gives us the global mass balance

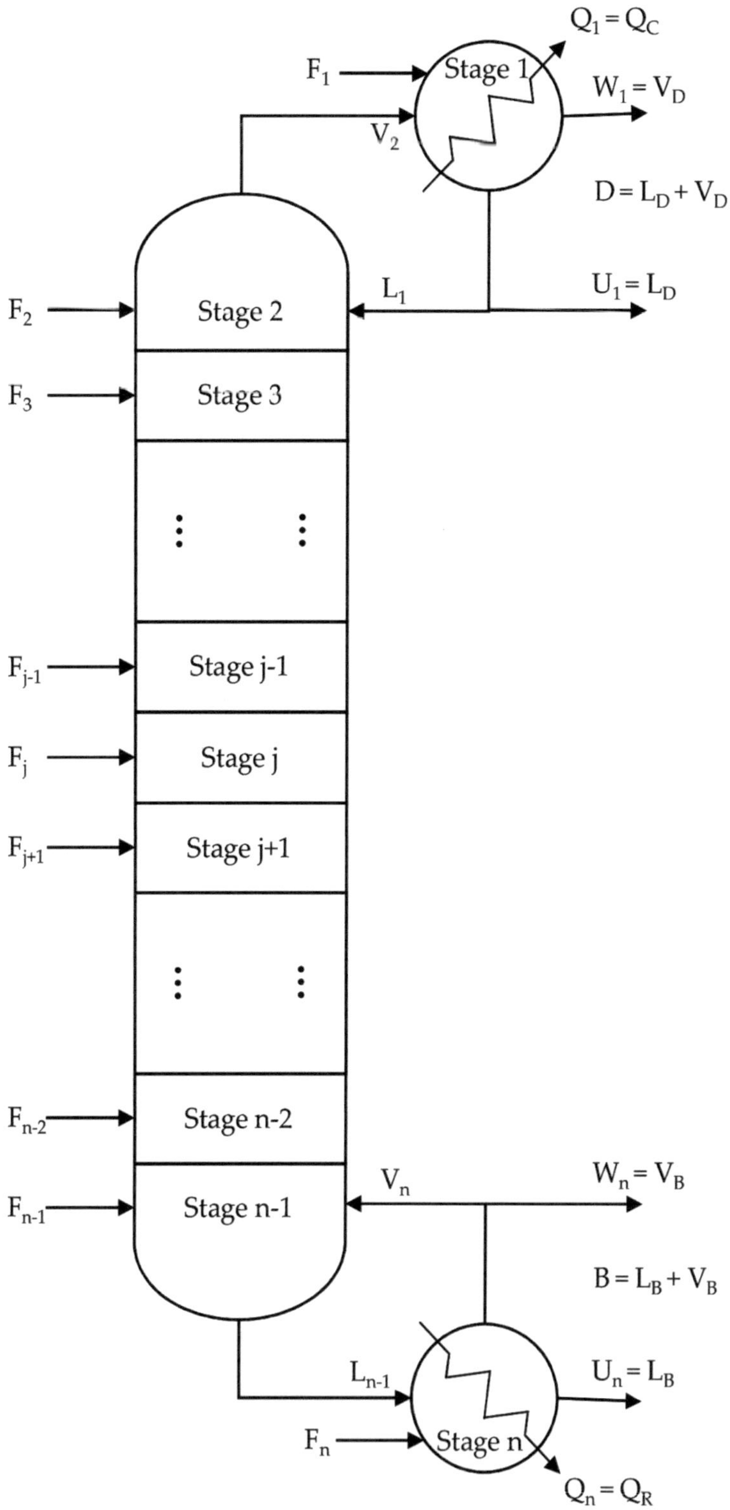

Figure 1. Schematic representation of a distillation column.

http://dx.doi.org/10.5772/66833

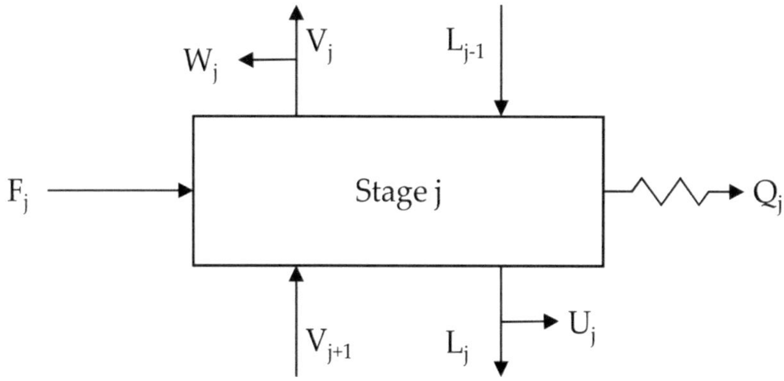

Figure 2. Schematic representation of input and output streams in a generic stage.

$$\sum_{j=1}^{n} F_j - \sum_{j=1}^{n} (W_j + U_j) = 0. \tag{2}$$

Or putting separated the bottoms and distillate (side streams in condenser and reboiler)

$$\sum_{j=1}^{n} F_j - B - D - \sum_{j=2}^{n-1} (W_j + U_j) = 0 \tag{3}$$

The mass balance of component *i* in stage *j* is given by

$$z_{ij} F_j + y_{i\,j+1} V_{j+1} + x_{i\,j-1} L_{j-1} - y_{ij} (V_j + W_j) - x_{ij} (L_j + U_j) = 0 \tag{4}$$

And the global mass balance of component *i* by

$$\sum_{j=1}^{n} z_{ij} F_j - \sum_{j=1}^{n} (y_{ij} W_j + x_{ij} U_j) = 0. \tag{5}$$

Or putting separated the bottoms and distillate

$$\sum_{j=1}^{n} z_{ij} F_j - x_{i\,n} L_B - y_{i\,n} V_B - x_{i\,1} L_D + y_{i\,1} V_D - \sum_{j=2}^{n-1} (x_{ij} W_j + x_{ij} U_j) = 0 \tag{6}$$

where x_{ij} is the fraction of component *i* in the liquid phase of stage *j*, y_{ij} is the fraction of component *i* in the vapor phase of stage *j* and z_{ij} is the fraction of component *i* in the feed stream of stage *j*.

The energy balance in stage *j*, ignoring changes in kinetic and potential energies, is given by

$$F_j H_j^F + V_{j+1} H_{j+1}^V + L_{j-1} H_{j-1}^L - (V_j + W_j) H_j^V - (L_j + U_j) H_j^L - Q_j = 0 \tag{7}$$

where H_j^F is the enthalpy of the feed stream to stage *j*, H_j^V is the enthalpy of the vapor stream from stage *j* and H_j^L is the enthalpy of the liquid stream from stage *j*.

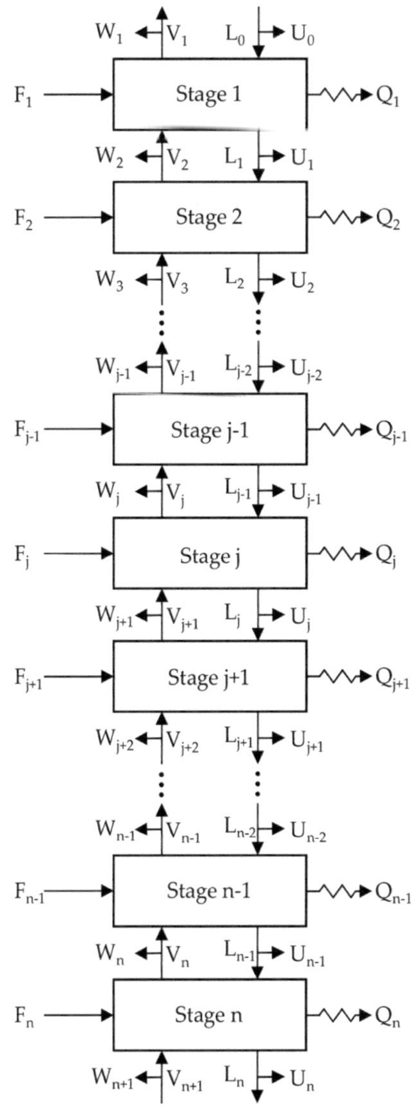

Figure 3. Schematic representation of input and output streams in all stages of a distillation column.

The global energy balance can be obtained by the sum of energy balance of all stages

$$\sum_{j=1}^{n} F_j H_j^F - \sum_{j=1}^{n} (W_j H_j^V + U_j H_j^L + Q_j) = 0 \tag{8}$$

Or in a equivalent form

$$\sum_{j=1}^{n} F_j H_j^F - L_B H_n^L - V_B H_n^V - L_D H_1^L - V_D H_1^V - Q_C - Q_R - \sum_{j=2}^{n-1} (W_j H_j^V + U_j H_j^L + Q_j) = 0 \tag{9}$$

Normally for the operation of a distillation column, are specified the reflux and the reboil ratios, respectively, defined by

$$r_D = \frac{L_1}{D} \tag{10}$$

$$r_B = \frac{V_n}{B} \tag{11}$$

To generalize this part will be defined d_j the ratio between the side stream and the stream outputting a stage and inputting the near stages, that is,

$$d_j = \frac{R_j}{L_j + V_j} = \frac{U_j + W_j}{L_j + V_j} \tag{12}$$

The side stream R_j is given by the sum of liquid side stream U_j and vapor side stream W_j, so it is also necessary to define the vapor fraction of side stream ω_j, that is,

$$W_j = \omega_j R_j \tag{13}$$

$$U_j = (1\text{–}\omega_j)\, R_j \tag{14}$$

Stage type	No side stream	Only liquid side stream	Only vapor side stream	Vapor and liquid side streams
Generic (stage *j*)	$d_j = 0$	$d_j \neq 0$	$d_j \neq 0$	$d_j \neq 0$
	$W_j = 0$	$\omega_j = 0$	$\omega_j = 1$	$0 < \omega_j < 1$
	$U_j = 0$	$R_j = d_j\,(L_j + V_j)$	$R_j = d_j\,(L_j + V_j)$	$R_j = d_j\,(L_j + V_j)$
		$W_j = 0$	$W_j = R_j$	$W_j = \omega_j\, R_j$
		$U_j = R_j$	$U_j = 0$	$U_j = (1\text{–}\omega_j)\, R_j$
Condenser (*j*=*1*)	Total reflux	Total condenser	Partial condenser	Partial condenser
	$d_1 = 0$	$d_1 = 1/r_D$	$d_1 = 1/r_D$	$d_1 = 1/r_D$
	$W_1 = V_D = 0$	$\omega_1 = \omega_D = 0$	$\omega_1 = \omega_D = 1$	$0 < \omega_1 = \omega_D < 1$
	$U_1 = L_D = 0$	$R_1 = d_1\, L_1 = D$	$R_1 = d_1\, L_1 = D$	$R_1 = d_1\, L_1 = D$
		$W_1 = V_D = 0$	$W_1 = V_D = R_1$	$W_1 = V_D = \omega_D\, R_1$
		$U_1 = L_D = R_1$	$U_1 = L_D = 0$	$U_1 = L_D = (1\text{–}\omega_D)\, R_1$
Reboiler (*j*=*n*)	Total reflux	Partial reboiler	Total reboiler	Partial reboiler
	$d_n = 0$	$d_n = 1/r_B$	$d_n = 1/r_B$	$d_n = 1/r_B$
	$W_n = V_B = 0$	$\omega_n = \omega_B = 0$	$\omega_n = \omega_B = 1$	$0 < \omega_n = \omega_B < 1$
	$U_n = L_B = 0$	$R_n = d_n\, V_n = B$	$R_n = d_n\, V_n = B$	$R_n = d_n\, V_n = B$
		$W_n = V_B = 0$	$W_n = V_B = R_n$	$W_n = V_B = \omega_B\, R_n$
		$U_n = L_B = R_n$	$U_n = L_B = 0$	$U_n = L_B = (1\text{–}\omega_B)\, R_n$

Table 1. Four types of side streams.

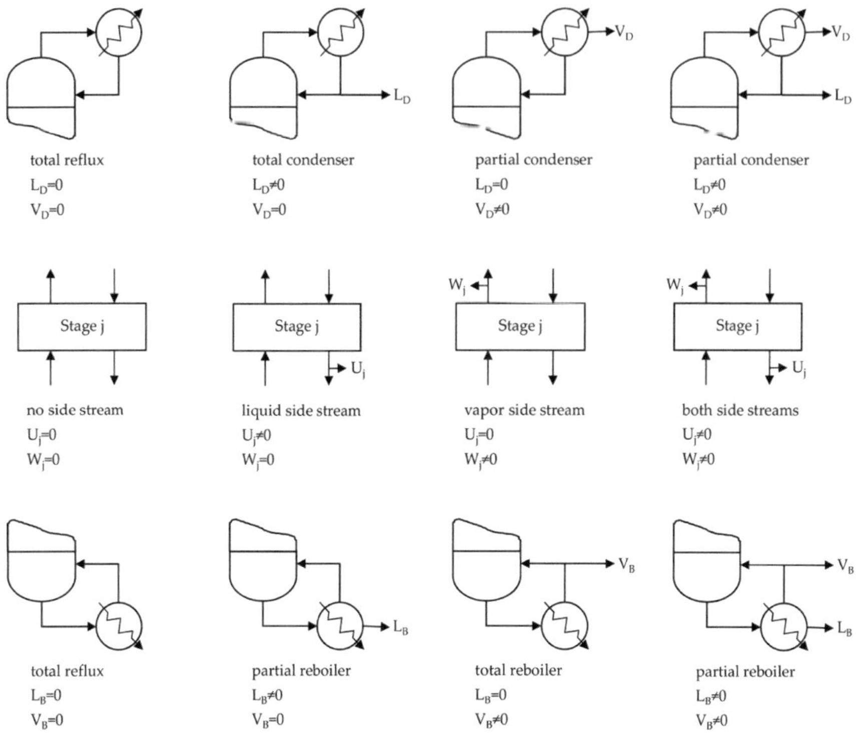

Figure 4. Four types of side streams.

How $V_1 = 0$, Eq. (12) is simplified to Eq. (10) by $d_1 = 1/r_D$ and how $L_n = 0$, Eq. (12) is simplified to Eq. (11) by $d_n = 1/r_B$.

Talking about side streams, there are four possibilities: no side stream, only liquid side stream, only vapor side stream and both liquid and vapor side stream. More details are shown in **Table 1** and **Figure 4**.

4. Thermodynamics

Thermodynamics plays a key role on the modeling of phase equilibrium. For the phases of liquid and vapor in thermodynamic equilibrium, the fraction of component in each phase is connected by the iso-fugacity relation [15–19]

$$f_i^L = f_i^V \tag{15}$$

where f_i^V is the fugacity of component i in vapor phase and f_i^L is the fugacity of component i in liquid phase, which leads to the mathematical relation

$$y_i = K_i\, x_i \tag{16}$$

where the way for calculating K_i depends on the chosen vapor liquid equilibrium (VLE) formulation, which can be Phi/Phi or Gamma/Phi [15–19], respectively,

$$K_i = \frac{\hat{\phi}_i^L}{\hat{\phi}_i^V} \tag{17}$$

$$K_i = \frac{\gamma_i\, \phi_i^{sat}\, P_i^{sat}}{\hat{\phi}_i^V} \exp\left(\int_{P_i^{sat}}^{P} \frac{V_i^L}{R\,T} dP\right) \tag{18}$$

where $\hat{\phi}_i^L$ is the fugacity coefficient of component i in liquid phase, $\hat{\phi}_i^V$ is the fugacity coefficient of component i in vapor phase, ϕ_i^{sat} is the fugacity coefficient of pure component i in saturation state, γ_i is the activity coefficient of component i in liquid phase, P is the pressure, P^{sat} is the saturated vapor pressure, R is the ideal gas constant, V_i^L is the component i volume in liquid phase and T is the temperature.

For the calculation of fugacity coefficient is required to use an equation of state (EOS) and for the calculation of activity coefficient is required to use a model that represents the excess Gibbs free energy. The saturated vapor pressure is calculated using one of the equation that describes the relation between vapor pressure and temperature for pure components, like Antoine, Wagner, Riedel, Harlecher-Braun, among others [15–19]. These equations are based on Clapeyron equation and their constant is obtained by experimental data fitting.

Another very important calculation that is needed to resort thermodynamics is the enthalpy. The liquid and vapor phases enthalpies are calculated [15–19], respectively, by

$$H^L = \sum_{i=1}^{m} x_i\, H_i^L + H^E \tag{19}$$

$$H^V = \sum_{i=1}^{m} y_i\, H_i^V + H^R \tag{20}$$

where m is the number of components, H_i is the enthalpy of pure component i, H^E is the excess enthalpy that can be calculated by a model that represents the excess Gibbs free energy (EOS or activity coefficient) and H^R is the residual enthalpy calculated using an EOS.

For the calculation of pure component i enthalpy, it is necessary to specify a reference state. How we are considering that there is no reaction in the column and it is possible to put the enthalpies of all components equal to zero in a reference state (same conditions of temperature, pressure and phase). But, the intention here is to make the model as general as possible it will be chosen elemental reference state that can be used also in a reactive system. In this reference state (298K and 1 atm), the standard enthalpy of formation (ΔH_f), by convention, for an

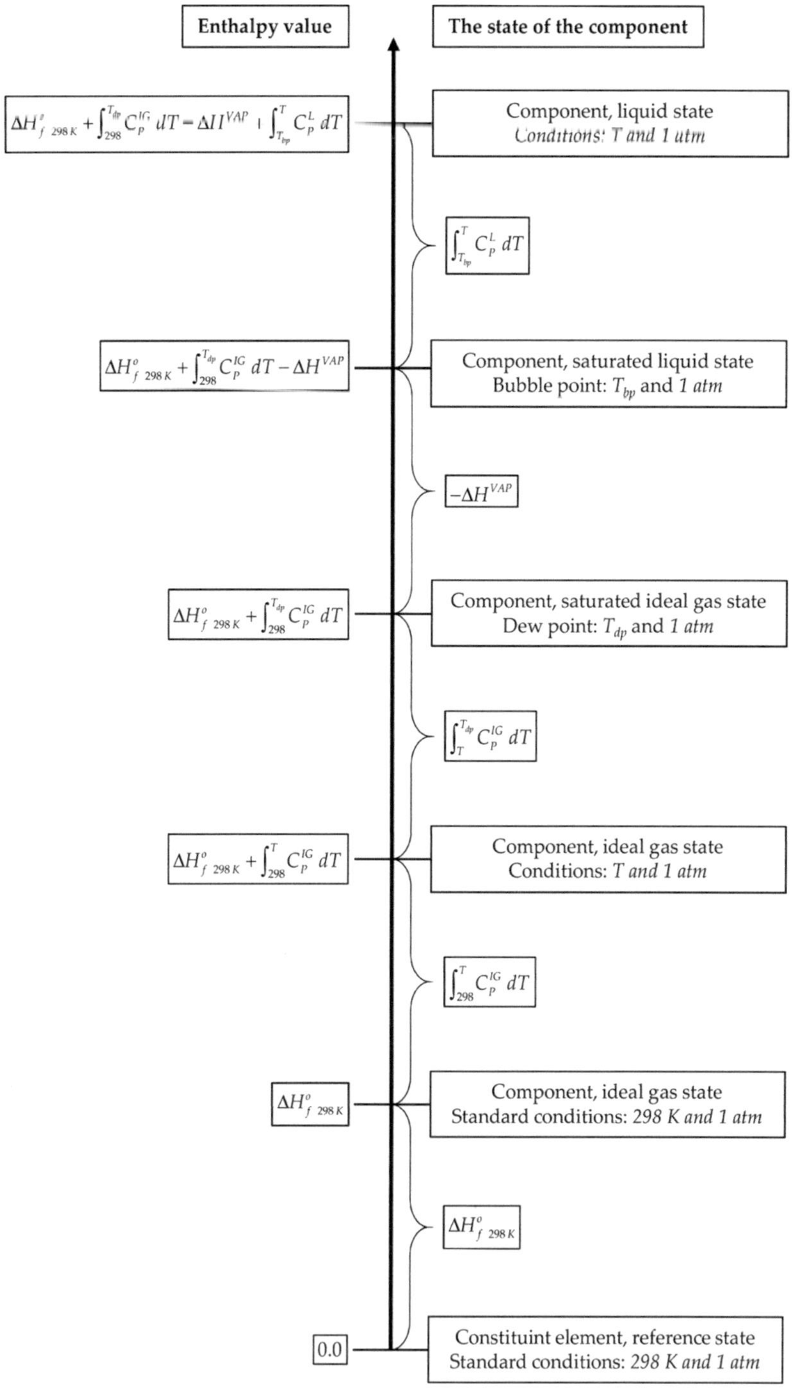

Figure 5. Calculation of the enthalpy of a component based on the elemental reference state.

element in its most stable form standard state is zero [15–19]. The calculation of the enthalpy of a component is based on the elemental reference state, shown in **Figure 5**.

Where C_P is the heat capacity at constant pressure of liquid (L) and ideal gas (IG), T_{bp} is the bubble point temperature, T_{dp} is the dew point temperature and ΔH^{VAP} is the vaporization enthalpy.

5. Materials and methods

In this section, will be presented the methods for generating the initial estimates and for solution of equations showed above.

5.1. Initial estimates

The resolution of equations that models the steady-state equilibrium distillation column involves a set of highly nonlinear equation, mainly on phase equilibria and energy balances. The algorithms, for solution of this type of problem, request good initial estimates in order that it can be possible to reach a solution. Moreover, here we are working in an algorithm that avoids the use of methods for solving nonlinear equations systems, what makes the quality of initial estimate even more important.

There are a lot of methods for solving models of steady-state equilibrium distillation column, considering various levels of layout complexity, number of components involved and accuracy of properties calculation. Simpler models do not need initial estimates, but for more complex models, a good initial estimate is fundamental.

McCabe-Thiele is a graphical method for combining the equilibrium curve with mass balance, assuming that there are two sections in the distillation column (between reboiler and feed stage and between feed stage and condenser) where molar vapor and liquid flow rates are constant, in addition to the assumption that there is no heat loss, eliminates the need of energy balances [4], something like the non-heat effect presented below. Ponchon-Savarit is a graphical method that includes energy balances, utilizing for this an enthalpy-concentration diagram [5, 6]. How, Ponchon-Savarit method utilizes energy balances and it is more accurate than McCabe-Thiele method. These methods do not need initial estimates but, unfortunately, are applicable only for distillation of binary mixtures.

For complex systems, it is suggested that the procedure for solving this models should be based on the solution of a system of nonlinear equations using an appropriated method for solving systems of nonlinear equations like Newton-Raphson. This system of equations is composed by MESH equations or combinations of them. But, the solution convergence of this type of problem is totally dependent on the quality of initial guess.

There are a lot of methods that use a technique called "tearing equations" that split large and sparse systems of algebraic equations into smaller system [20]. They are relatively simple, but are restricted to ideal and nearly ideal mixtures. The methods of Lewis-Matheson [9], Thiele-Geddes

[10] and theta [8] are based on equation tearing for solving simple distillation columns with one feed and two product stream. The bubble-point method receives this name because it tears the MESH equations in a way that a new set of stage temperatures is computed from bubble-point equations [11]. Similarly, the sum-rates method calculates, at each new iteration, the values of liquid streams by the summation of components flow rates in liquid phase [7].

The MESH equations wrote in this work will be rearranged for using tearing-equation method like bubble-point method, as can be seen in the next section. So, the idea here is to propose a method for generating good initial guesses aiming to avoid instability, normally presented for simpler method. That is, we are trying to join the efficiency of Newton-Raphson-like methods with the simplicity of methods based on tearing equations. It is already demonstrated that, with good initial guesses, it is possible to use tearing equation for very complex models like steady state for reactive distillation columns [13].

The algorithm at issue needs initial estimates for temperature, liquid and vapor streams and side streams. For initial estimates of temperature will be considered a model based on saturation temperature (T^{SAT}) of pure components at column operation pressure (P_{col}) and liquid, vapor and side streams will be estimated considering a non-heat effect.

For the initial estimates of stage temperatures ($T^{(0)}$) will be considered a linear profile, in which, approximately in the middle of the column, we have an average temperature (T_{ave}) pondered on component fractions in feed streams

$$T_{ave} = \frac{\sum_{i=1}^{m} \left(T_i^{SAT} \sum_{j=1}^{n} z_{ij} F_j \right)}{\sum_{j=1}^{n} F_j} \tag{21}$$

And a minimal temperature equal to T_{ave} minus the average absolute deviation (AAD)

$$T_{\min} = T_{ave} - AAD = T_{ave} - \frac{\sum_{i=1}^{m} \left(|T_i^{SAT} - T_{ave}| \sum_{j=1}^{n} z_{ij} F_j \right)}{\sum_{j=1}^{n} F_j} \tag{22}$$

That is,

$$T_j^{(0)} = T_{\min} + \frac{2\,(j-1)}{n}\,(T_{ave} - T_{\min}) \tag{23}$$

For non-heat effects model is made the assumption of constant vaporization enthalpy, what can guarantee that the change in the liquid and vapor flow rates in each section of the column is due to only the feed quality, feed flow rate and side flow rates [21], as illustrated in **Figure 6**. Where the feed quality is defined as

- Saturated liquid;

http://dx.doi.org/10.5772/66833

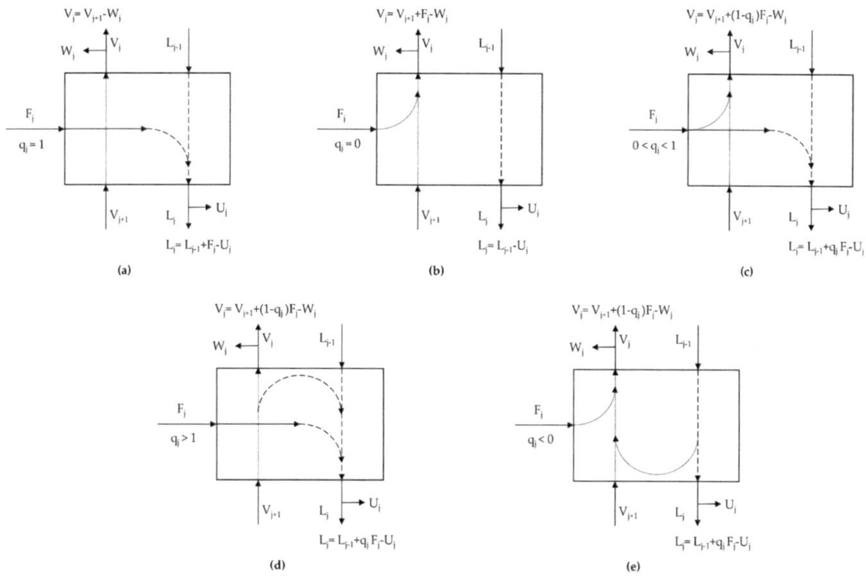

Figure 6. Non-heat effect behavior in a stage for: (a) saturated liquid; (b) saturated vapor; (c) saturated liquid and vapor; (d) subcooled liquid and (e) superheated vapor.

$$q = 1 \tag{24}$$

- Saturated vapor;

$$q = 0 \tag{25}$$

- Saturated liquid and vapor, q is equal to liquid fraction;

$$0 < q < 1 \tag{26}$$

- Subcooled liquid, $q > 1$;

$$q = \left(\Delta H^{VAP} + \int_{T_{bp}}^{T} C_P^L \, dT\right) \Big/ \Delta H^{VAP} \tag{27}$$

- Superheated vapor, $q < 0$.

$$q = -\left(\int_{T_{dp}}^{T} C_P^V \, dT\right) \Big/ \Delta H^{VAP} \tag{28}$$

Using the mass balance in a stage and the global mass balance, one can find the expression

$$R_j = \frac{\frac{d_j}{d_j+1}\left(F_j + V_{j+1} + L_{j-1}\right)}{\sum\limits_{j=1}^{n} \frac{d_j}{d_j+1}\left(F_j + V_{j+1} + L_{j-1}\right)} \sum_{j=1}^{n} F_j \tag{29}$$

At this point, we want to know just a rude value of side flow rates. So, let us make the rude assumption that feed and side streams are 50% in each phase and assume that the non-heat effects are valid. In this case, the expression

$$\frac{1}{d_j+1}\left(F_j+V_{j+1}+L_{j-1}\right) \tag{30}$$

is almost constant along the column. And, it can be used to generate initial estimates for side flow rates

$$R_j^{(0)}=\frac{d_j}{\sum\limits_{j=1}^{n} d_j}\sum_{j=1}^{n} F_j \tag{31}$$

If there are side streams only in the reboiler and in the condenser or the side streams in the intermediate stages are known, the non-heat effects model can be used to generate initial estimates directly to bottoms and distillate flow rates

$$B^{(0)}=\frac{r_D\,F_1+(r_D+1)\,F_n+\sum\limits_{j=2}^{n-1}\left[(q_j+r_D)\,F_j-(r_D+1)\,U_j-W_j\right]}{r_D+r_B+1} \tag{32}$$

$$D^{(0)}=\frac{(r_B+1)\,F_1+r_D\,F_n+\sum\limits_{j=2}^{n-1}\left[(r_B+1-q_j)\,F_j+(r_D+1)\,U_j+W_j\right]}{r_D+r_B+1} \tag{33}$$

So V_n and L_1 can be estimated using, respectively, the reboil and reflux ratio

$$V_n^{(0)}=r_B\,B^{(0)} \tag{34}$$

$$L_1^{(0)}=r_D\,D^{(0)} \tag{35}$$

Using the non-heat effects model, it also can be obtained the initial estimates for vapor flow rates, by mass balances in vapor phase, for $j=n-1,\ \ldots,2$

$$V_j^{(0)}=\frac{(1-q_j)\,F_j+V_{j+1}^{(0)}}{d_j+1} \tag{36}$$

The liquid flow rates are estimated using the mass balance by stage for $j=2,\ \ldots,n-1$

$$L_j^{(0)}=\frac{F_j+V_{j+1}^{(0)}+L_{j-1}^{(0)}}{d_j+1}-V_j^{(0)} \tag{37}$$

With the initial estimates for side, liquid and vapor stream and temperature, it can be started the iterative process of steady-state model solution. The only unknowns lasting, to generate initial estimates, are the components fractions in each stream, but, in the algorithm, we are going to work; these initial estimates, showed how to generate, are sufficient for the calculation of components fractions. So, for these unknowns, it is not necessary to generate initial estimates.

5.2. Algorithm

When one talks about the solution of algebraic equation, it is necessary to keep in mind that for the calculation of a number unknowns is necessary the same number of equations. In a complex system of equations, it is not easy to be sure that we have the right number of equations and unknowns. To make it easier to do, this balance is presented in **Table 2**.

Unknown type	Number of unknowns	Stage	Equation(s) used to calculate
Temperatures	n	1 to n	The restriction that the sum of vapor mole fraction is equal to unity, for stages 1 to n
Vapor flow rates	n–1	2 to n	Energy balance on stages to n–1 and mass balance on stage n
Liquid flow rates	n–1	1 to n–1	Mass balance for stage 1 to n–1
Side flow rates (liquid and vapor)	2 n	1 to n	Ratio between side streams and streams outputting stages for n stages. And vapor fraction of side stream for n stages
Liquid mole fractions	$n\ m$	1 to n	Mass balance by component in n stages for m components
Vapor mole fractions	$n\ m$	1 to n	Iso-fugacity relation for m components in n stages
Heat transfer	2	1 and n	Energy balance in 1 stages and n
Total	$(2\ m + 5)\ n$	n	$(2\ m + 5)\ n$

Table 2. Balance of equations and unknowns.

The first step of algorithm is the imputation data. The data needed to be imputed are as follows:

- Number of components, m;
- Number of stages, n;
- The side streams ratios, d_j for $j = 1, \ldots, n$ ($d_n = 1/r_B$ and $d_1 = 1/r_D$);
- The vapor fraction of side stream, ω_j for $j = 1, \ldots, n$;
- Column pressure, P_{col};
- The heat transfer for intermediate stages, Q_j for $j = 2, \ldots, n$–1(they are normally adiabatic stages, that is, $Q_j = 0$);
- Feed flow rates, F_j for $j = 1, \ldots, n$;
- Feed temperature, T_j^{Feed} for $j = 1, \ldots, n$;

- Feed quality, q_j for $j = 1, \ldots, n$;
- Feed fractions, $z_{i\,j}$ for $j = 1, \ldots, n$ and $i = 1, \ldots, m$;
- Physical, critical and other properties of components, for evaluation of enthalpies and phase equilibrium.

At this point, a more watchful reader must be thinking: it was used the restriction of vapor mole fraction sum, but was not used the same restriction for the liquid mole fraction, that is, right? Actually, this restriction is implicitly used, because the mass balance of a stage is the sum of mass balances by component in that stage and the sum of liquid mole fractions is forced to be equal to unity by a normalization step used in the algorithm presented ahead.

The sequence of calculation of the algorithm is presented in **Figure 7**.

Using the mass balance by stage and the phase equilibrium relation

$$y_{i\,j} = K_{i\,j}\, x_{i\,j} \tag{38}$$

One obtain

$$l_j\, x_{i\,j-1} + m_j\, x_{i\,j} + u_j\, x_{i\,j} = b_j \tag{39}$$

where

$$l_j = L_{j-1} \tag{40}$$

$$a_j = -\left[(L_j + U_j) + (V_j + W_j)\, K_{i\,j}\right] \tag{41}$$

$$u_j = V_{j+1}\, K_{i\,j+1} \tag{42}$$

$$b_j = -F_j\, z_{i\,j} \tag{43}$$

Remembering that the L_0, L_n, V_1 and V_{n+1} streams do not exist, it is obtained a tri-diagonal linear system of equations for each component

$$\begin{bmatrix} a_1 & u_1 & 0 & 0 & 0 \\ l_2 & a_2 & u_2 & 0 & 0 \\ 0 & \ddots & \ddots & \ddots & 0 \\ 0 & 0 & l_{n-1} & a_{n-1} & u_{n-1} \\ 0 & 0 & 0 & l_n & a_n \end{bmatrix} \cdot \begin{bmatrix} x_{i\,1} \\ x_{i\,2} \\ \vdots \\ x_{i\,n} \\ x_{i\,n} \end{bmatrix} = \begin{bmatrix} b_1 \\ b_2 \\ \vdots \\ b_n \\ b_n \end{bmatrix} \tag{44}$$

Because the initial guess imprecision (mainly on temperatures) and the mole fractions of each component are calculated separately, especially in the first iterations, they may have values without any physical meaning (such as, negative or a sum different from unity,). So that, the convergence process of algorithm may be accelerated by normalization, undertaken by the following equation

http://dx.doi.org/10.5772/66833

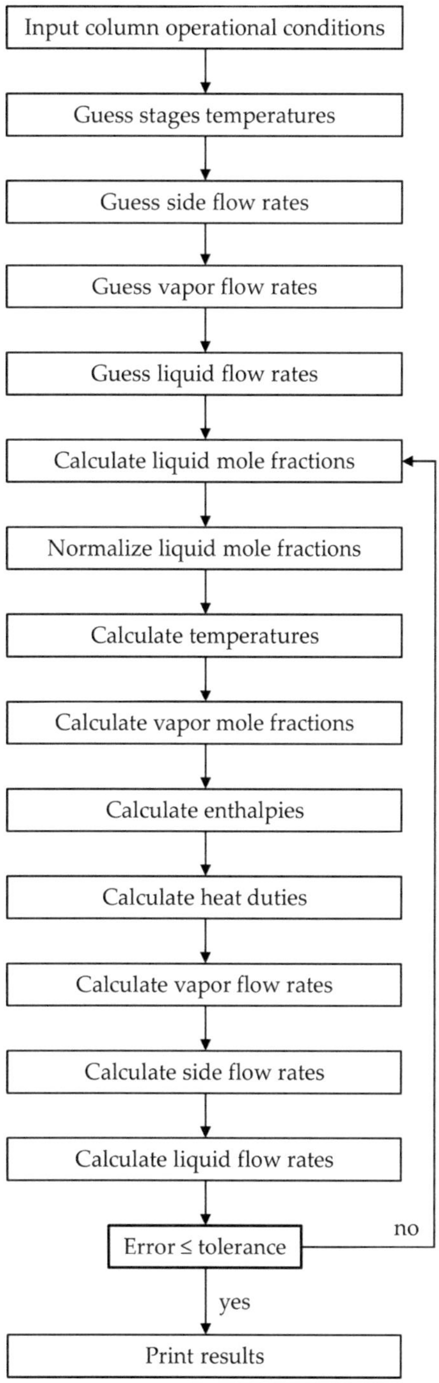

Figure 7. Algorithm for the simulation of steady-state distillation columns.

$$x_{ij} = \frac{|x_{ij}|}{\sum_{i=1}^{m} |x_{ij}|} \tag{45}$$

The temperatures are the only set of unknowns that cannot be calculated by a linear method. The temperatures are calculated stage by stage using the restriction that the sum of mole fractions in vapor phase must be equal to unit with aid of phase equilibrium relation, that is,

$$F(T_j) = 1 - \sum_{i=1}^{m} K_{ij}\, x_{ij} = 0 \tag{46}$$

The solution of this step is made with aid of a iterative root-finding method, like secant

$$T_j^{k+1} = T_j^k - F(T_j^k) \left[\frac{T_j^k - T_j^{k-1}}{F(T_j^k) - F(T_j^{k-1})} \right] \tag{47}$$

With the values of temperature and liquid mole fractions, it can be calculated the vapor mole fractions simply by using the phase equilibrium relation, that is, Eq. (38).

The enthalpies of the all streams are calculated as previously shown. These enthalpies are used in the energy balances. The energy balances in stages 1 and n are used to calculate the heat duties on these stages. In the others stages, the heat transfer must be zero (adiabatic stages) or must be fixed in the input data.

$$Q_C = Q_1 = F_1\, H_1^F + V_2\, H_2^V - W_1\, H_1^V - (L_1 + U_1)\, H_1^L \tag{48}$$

$$Q_R = Q_n = F_n\, H_n^F + L_{n-1}\, H_{n-1}^L - (V_n + W_n)\, H_n^V - U_n\, H_n^L \tag{49}$$

The value of vapor flow rate from stage n is obtained by using the mass balance in this stage

$$V_n = F_n + L_{n-1} - U_n - W_n \tag{50}$$

And the values of vapor flow rate from stages 2, …, n−1 are calculated using energy balance with the aid of mass balances to eliminate the liquid flow rates terms. The calculation is made from n−1 back to 2

$$V_j = \frac{\beta_j\, V_{j+1} + \gamma_j}{\alpha_j} \tag{51}$$

where,

$$\gamma_j = (H_j^F - H_j^L)\, F_j + (H_{j-1}^L - H_j^L) \sum_{k=1}^{j-1} (F_k - U_k - W_k) + (H_j^L - H_j^V) W_j - Q_j \tag{52}$$

$$\beta_j = H_{j+1}^V - H_j^L \tag{53}$$

http://dx.doi.org/10.5772/66833

$$\alpha_j = H_j^V - H_{j-1}^L \tag{54}$$

The energy balance normally is used to calculate the temperatures, but it would be necessary a solution of a system of highly nonlinear equations. Instead of this, the energy balances are used to calculate the vapor flow rates, with aid of mass balances and some algebraic rearrangement, by sequential evaluations (one vapor flow rate at a time).

The side flow rates are calculated by substituting the ratio between the side stream and the stream outputting a stage in the mass balance

$$R_j = \left(\frac{d_j}{d_j + 1}\right)(F_j + V_{j+1} + L_{j-1}) \tag{55}$$

$$W_j = \omega_j\, R_j \tag{56}$$

$$U_j = (1 - \omega_j)\, R_j \tag{57}$$

The liquid flow rates are calculated sequentially from stage 1 to stage $n-1$(remember, there is no L_n) by using the mass balance in the respective stages

$$L_j = F_j + V_{j+1} - V_j + L_{j-1} - R_j \tag{58}$$

The iterative process finishes when the relative variation of some main unknowns of the model is very low. The smallness of the variation depends on the chosen tolerance

$$\sum_{j=1}^{n}\left[\frac{T_j^{(k)} - T_j^{(k-1)}}{T_j^{(k)}}\right]^2 + \sum_{j=1}^{n-1}\left[\frac{L_j^{(k)} - L_j^{(k-1)}}{L_j^{(k)}}\right]^2 + \sum_{j=2}^{n}\left[\frac{V_j^{(k)} - V_j^{(k-1)}}{V_j^{(k)}}\right]^2 \le tolerance \tag{59}$$

6. Results and discussion: case study

For the evaluation of the initial estimates and algorithm in question, it will be tested with an example of hydrocarbons separation. Where a feed stream contains four hydrocarbons: propane (C_3), n-butane(n-C_4), isopentane (i-C_5) and n-pentane (n-C_5). The operational conditions are presented in **Table 3**.

For representing the nonideal behavior of vapor phase, it was used Peng-Robinson equation of state, where were considered as mixing rule binary iteration parameters with geometric mean for parameter a and arithmetic mean for parameter b. And for representing the liquid non-ideality, it was utilized UNIFAC method. The UNIFAC parameters, physical and critical properties for the components involved in the case of study, were taken from [16, 22], in which physical and critical properties are presented in **Table 4**.

$$Cp = C_1 + C_2\,T + C_3\,T^2 + C_4\,T^3 \tag{60}$$

Variables	Specifications	
Pressure	All stages	13.8bar
Ratio	Reflux	r_D =5.0
	Reboil	r_B =3.2531
Stages	Number	12
	Type	1 condenser 10 adiabatic stages ($Q_j = 0$) 1 reboiler
Feed	Location	Stage 6
	Condition	Saturated liquid
	Flow rate	100 kmol/h
	Temperature	359.3K
	Mole fractions	C_3 (0.4)
		n-C_4 (0.4)
		i-C_5 (0.1)
		n-C_5 (0.1)
Side streams	Condenser (stage 1)	$d_1 = 1/r_D$
		$\omega_1 = \omega_D = 0$
		$D = L_D$
	Stages 2–11	$d_j = 0$
		$U_j = W_j = 0$
	Reboiler (stage 12)	$d_{12} = 1/r_B$
		$\omega_{12} = \omega_B = 0$
		$B = L_B$

Table 3. Operational conditions of distillation column.

$$\ln(P^{sat}) = A - \frac{B}{T + C} \tag{61}$$

For a tolerance of 1×10^{-10}, it was needed 28 iterations to achieve convergence. A graphic of error versus iteration is shown in **Figure 8**. In the start of the procedure, few steps present a high error reduction, after this a linear convergence rate is obtained. That is not as good as a quadratic convergence of an algorithm that uses a method for solution of nonlinear equations, like Newton-Raphson, but is justified by the simplicity of the solution algorithm.

Figure 9 shows a comparison between initial guess and simulation results for stage temperatures. Initial guess profile is extremely close to results.

Figure 10 shows comparisons between initial guess and simulation results for molar flow rates of the liquid and vapor phases. Initial guess of rates of both phases was extremely close to the values calculated by simulation.

Properties		Propane	n-Butane	isoPentane	n-Pentane
Properties for calculation of fugacity coefficients	T_c (K)	369.8	425.2	460.4	469.7
	P_c (bar)	42.5	38.0	33.9	33.7
	ω	0.153	0.199	0.227	0.251
	T_{bp} (K)	231.1	272.7	301.0	309.2
	V^L (m^3/kmol)	0.0758	0.1004	0.1164	0.1152
C_P^{IG} (kJ/(kmol K)) T (K)	C_1	-4.224	9.487	-9.525	-3.626
	C_2	3.063	3.313	5.066	4.873
	C_3	-1.586	-1.108	-2.729	-2.580
	C_4	3.215	-0.2822	5.723	5.305
C_P^L (kJ/(kmol K)) T (K)	C_1	87.19	110.4	115.6	133.7
	C_2	-0.6110	-0.6567	-0.1138	-0.7024
	C_3	4.128	5.016	5.773	5.904
P^{sat} T (K) and P (bar)	A	9.1058	9.0580	9.0136	9.2173
	B	1872.46	2154.90	2348.67	2477.07
	C	-25.16	-34.42	-40.05	-39.94

Table 4. Physical and critical properties for the components involved in the case study.

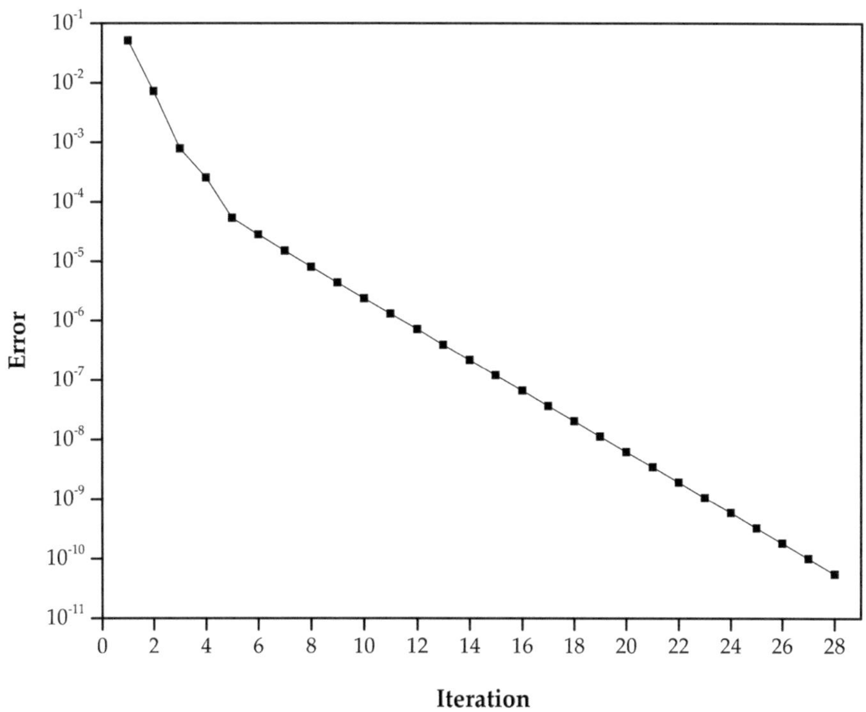

Figure 8. Numerical convergence.

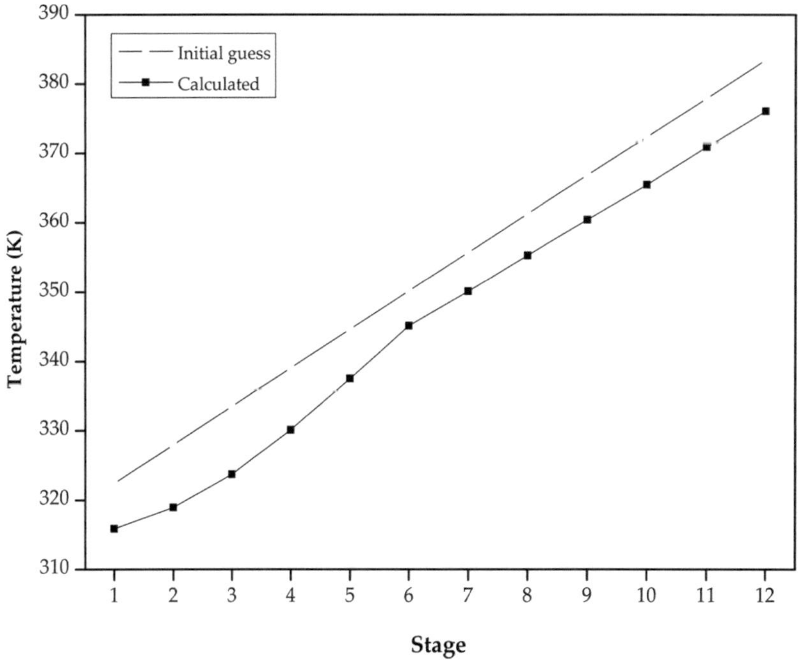

Figure 9. Temperature profiles.

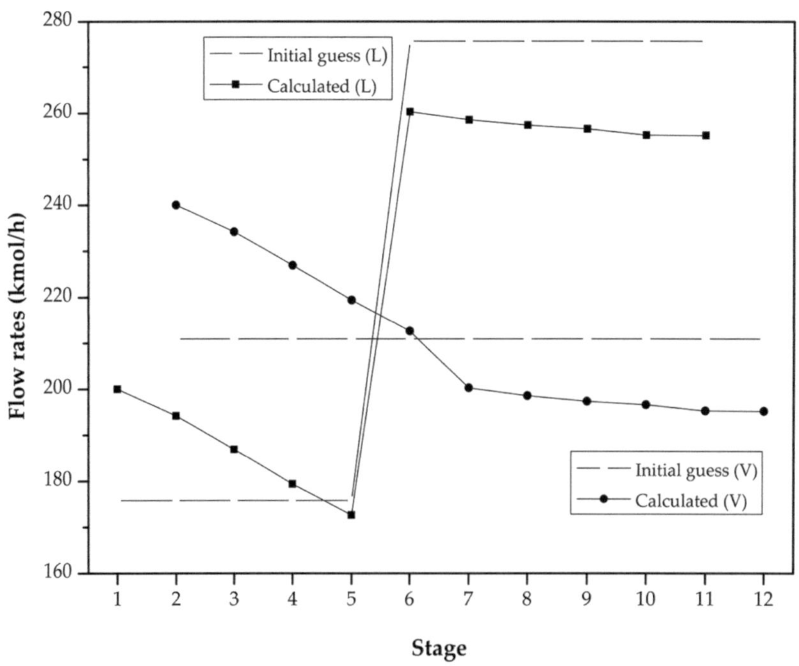

Figure 10. Liquid and vapor flow rates profiles.

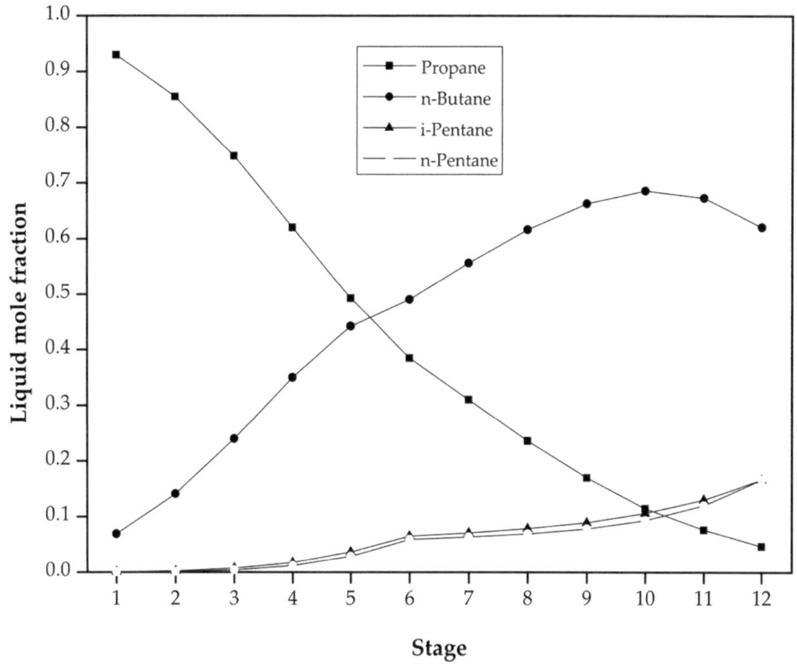

Figure 11. Liquid mole fractions profiles.

		Value	
Variables	**Specification**	**This work**	**King [23]**
Flow rate (kmol/h)	D	40.000	40.0
	B	60.000	60.0
Heat duty (kJ/h)	Q_C	4.1133×10^6	–
	Q_R	-4.0774×10^6	–
Temperature (K)	Condenser	315.91	316.15
	Reboiler	376.10	377.15
Distillate mole fractions	C_3	0.92966	0.980
	n-C_4	0.06936	0.020
	i-C_5	0.00070	0.000
	n-C_5	0.00029	0.000
Bottoms mole fractions	C_3	0.04688	0.013
	n-C_4	0.62044	0.653
	i-C_5	0.16621	0.167
	n-C_5	0.16648	0.167

Table 5. Some simulation results.

The liquid mole fraction profiles calculated are shown in **Figure 11**. Some important numeric results of the simulation are shown in **Table 5**.

In this case of study, the initial estimates generated are no more than 20% far from the final result, what confirm the goodness of the methodology used for generating the guesses. The results obtained here are very close to that obtained by King [23], some differences can be justified by different levels of accuracy of the thermodynamic modeling (the thermodynamic modeling of the cited reference is simpler).

This case being studied aims to separate the propane from the other three hydrocarbons. For evaluating the influence of feed stage is presented in **Figure 12**, which presents the fraction of propane in distillate stream in function of the stage where the feed stream occurs. One can see in this figure that the best stage to put the feed stream is in the middle of the column. It is easy to understand, if the feed occurs near of condenser, there will be a great amount of components other than propane in the stages near of condenser, so part of it eventually outputs the column in the distillate stream and if the feed occurs near of reboiler, there will be a great amount of propane in the stages near of reboiler, so part of propane is present on bottoms stream.

The number of stage also has a great influence on the fraction of propane in distillate stream. It is obvious that, the more stages there are in the column, a greater mole fraction of propane there will be in distillate. But, it is possible to see in **Figure 13** that after a certain number of stages, the increase on that fraction is too small. **Figure 12** confirms what was verified in

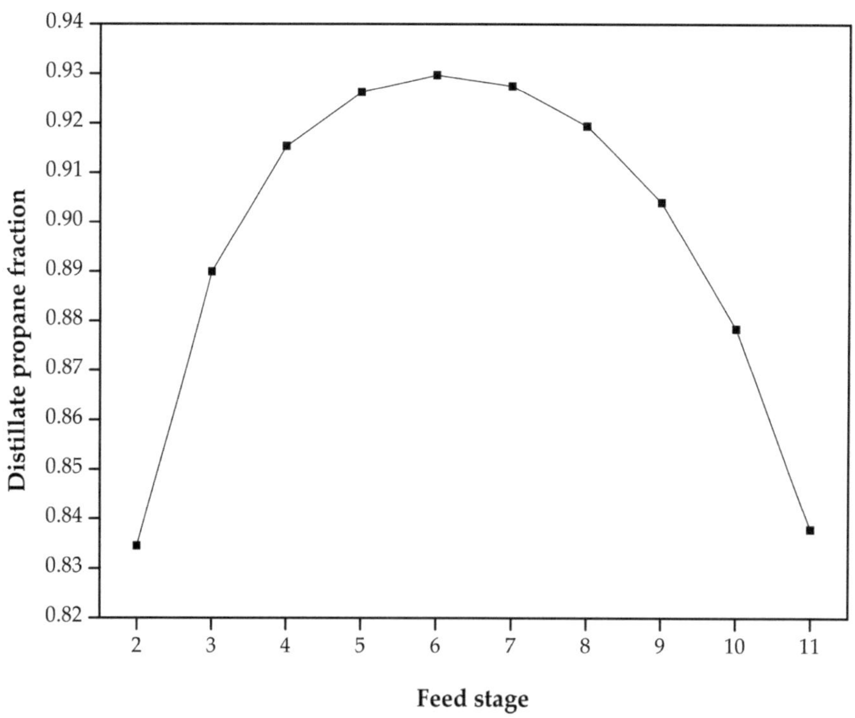

Figure 12. Fraction of propane in distillate stream in function of feed stage.

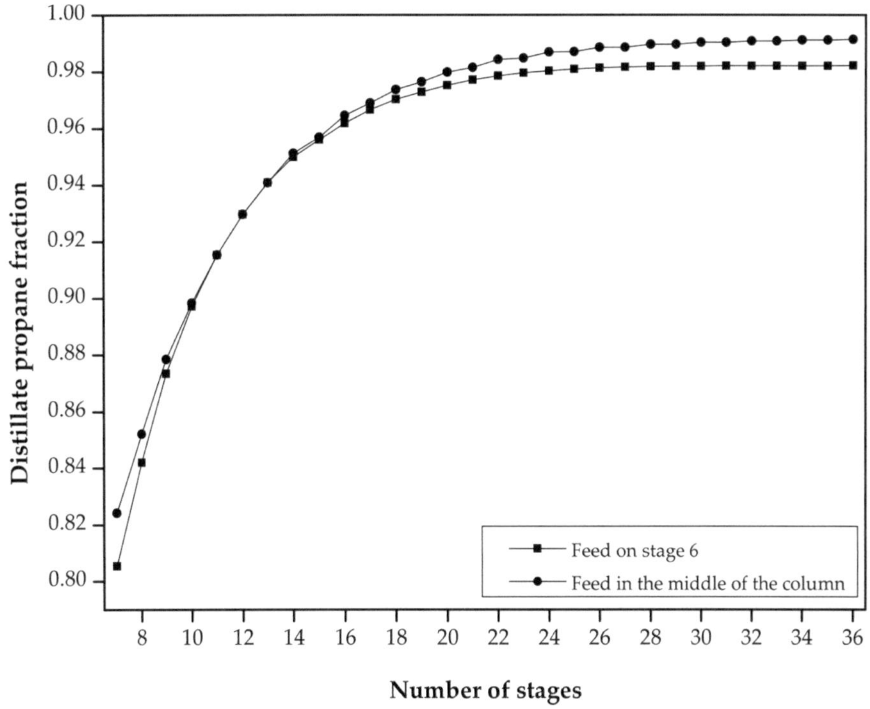

Figure 13. Fraction of propane in distillate stream in function of the number of stages.

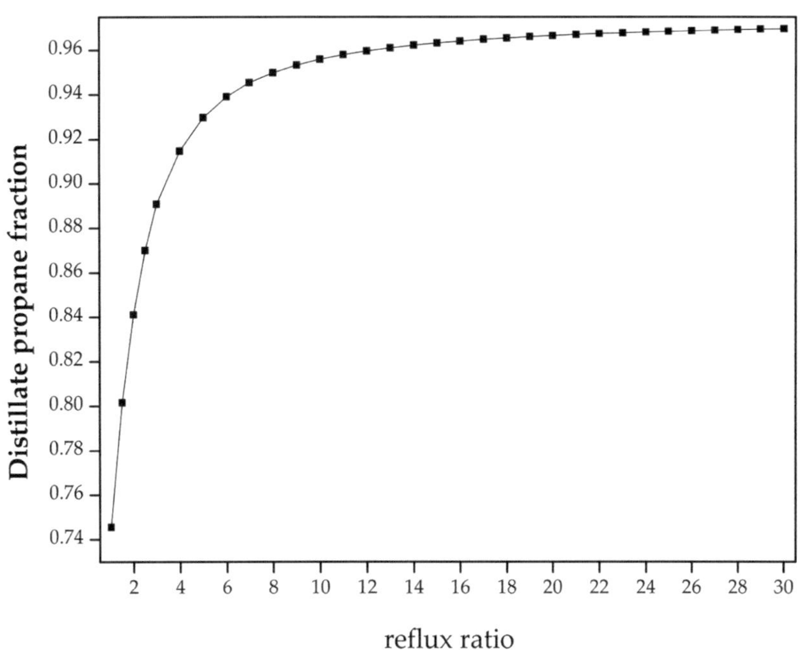

Figure 14. Fraction of propane in distillate stream in function of the reflux ratio.

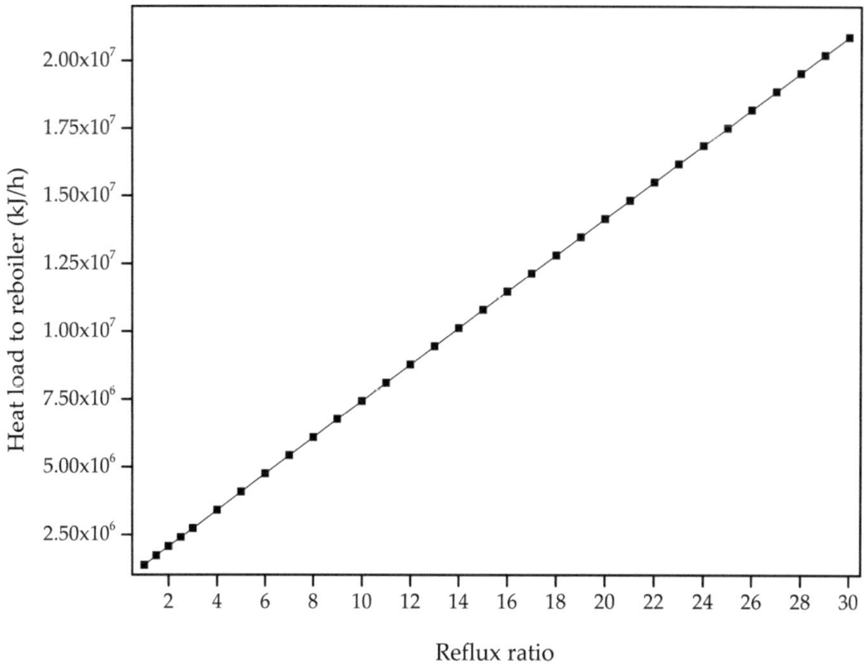

Figure 15. Heat load to reboiler in function of the reflux ratio.

Figure 13, because there is a comparison between column with the same number of stage, one with feed stream in stage six and another in the middle of column. And again the feed in the middle of column presents a better separation.

Another parameter that has a strong influence on the mole fraction of propane in distillate stream is the reflux ratio, see **Figure 14**. Higher purity levels are achieved by increasing the reflux ratio, with a clear limit. But greater reflux ratio greater is the heat load to the reboiler, see **Figure 15**. After a certain value of reflux ratio, the increase in propane mole fraction is very small.

With everything that has been exposed, one can see that achieving a better separation level is result mainly by spending more in operating or investment costs (greater reflux ratio, more stages, etc.) or changing the layout configuration of the operating system (feed stage location).

7. Conclusion

An algorithm was provided in this work for the solution of a distillation column operating in steady state; in this algorithm, the high nonlinear equations are solved in a very simple form. Equations in the model were divided into sets and each set was solved separately. The solution procedure uses an algorithm for solution of systems of tri-diagonal linear equation, explicit calculations and a method for root finding of equations of one unknown variable. A methodology was also provided to produce initial guess which constitutes a critical step in the solution of nonlinear equations system. The modeling allows a variety of cases depending on the types of condenser and reboiler, number and conditions of feed stream, side streams, etc.

The suggested methodology for the production of initial estimates was efficient with values close to those calculated by simulation. This fact accelerates and increases the convergence warranty.

Author details

Vilmar Steffen[1]* and Edson Antonio da Silva[2]

*Address all correspondence to: vilmars@utfpr.edu.br

1 Academic Department of Chemical Engineering (DAENQ), Federal University of Technology – Paraná (UTFPR), Francisco Beltrão, Brazil

2 Center for Engineering and Exact Sciences (CECE), State University of Western Paraná (Unioeste), Toledo, Brazil

References

[1] Marquini MF, Mariani DC, Meireles AJA, Santos OAA, Jorge LMM. Simulation and analysis of an industrial system of columns for ethanol distillation. Acta Scientiarum. Technology. 2007;**29**(1):23–28. doi:10.4025/actascitechnol.v29i1.81

[2] Seader JD, Henley EJ, Roper DK. Separation process principles: chemical and biochemical operations. 3rd ed. New York: John Wiley & Sons; 2011. 821 p.

[3] Corazza FC, Oliveira JV, Corazza ML. Application of a subdivision algorithm for solving nonlinear algebraic systems. Acta Scientiarum. Technology. 2008;**30**(1):27–38. doi:10.4025/actascitechnol.v30i1.3198

[4] McCabe WL, Thiele EW. Graphical Design of Fractionating Columns. Industrial & Engineering Chemistry. 1925;**17**(6):605–611. doi:10.1021/ie50186a023

[5] Ponchon M. Application of Graphs to the Study of Commercial Fractional Distillation. La Technique Moderne. 1921;**13**:20, 55.

[6] Savarit R. Arts et Metiers. 1922:65, 142, 178, 241, 266, 307.

[7] Burningham DW, Otto FD. Which computer design for absorbers. Hydrocarbon Processing. 1967;**46**(10):163–170.

[8] Holland CD. Fundamentals of Multicomponent Distillation. New York: McGraw-Hill; 1980. 626 p.

[9] Lewis WK, Matheson GL. Studies in distillation design of rectifying columns for natural and refinery gasoline. Industrial & Engineering Chemistry. 1932;**24**:496–498.

[10] Thiele EW, Geddes RL. Computation of distillation apparatus for hydrocarbon mixtures. Industrial & Engineering Chemistry. 1933;**25**(3):289–295. doi:10.1021/ie50279a011

[11] Wang JC, Henke GE. Tridiagonal matrix for distillation. Hydrocarbon Process. 1966;**45** (8):155–163.

[12] Seader JD, Henley EJ. Separations Process Principles. New York: John Wiley & Sons; 1998. 886 p.

[13] Steffen V, Silva EA. Steady-state modeling of reactive distillation columns. Acta Scientiarum. Technology. 2012;**34**(1):61–69. doi:10.4025/actascitechnol.v34i1.9535

[14] Alfradique MF, Castier M. Modeling and simulation of reactive distillation columns using computer algebra. Computers and Chemical Engineering. 2005;**29**(9):1875–1884. doi:10.1016/j.compchemeng.2005.04.002

[15] Sandler SI. Chemical, Biochemical and Engineering Thermodynamics. 4th ed. New York: John Wiley & Sons, Inc.; 2006. 945 p.

[16] Poling BE, Prausnitz JM, O'Connell JP. The Properties of Gases and Liquids. 5th ed. New York: McGraw-Hill; 2004.

[17] Smith JM, Van Ness HC, Abbott MM. Introduction to Chemical Engineering Thermodynamics. 7th ed. New York: MacGraw Hill; 2005. 817 p.

[18] Koretsky MD. Engineering and Chemical Thermodynamics. 2nd ed. New York: John Wiley & Sons, Inc.; 2013. 704 p.

[19] Elliott JR, Lira CT. Introductory Chemical Engineering Thermodynamics. 2nd ed. New York: Prentice Hall; 2012. 912 p.

[20] Krong G. The Piecewise Solution of Large-Scale Systems Hardcover. London: MacDonald & Co; 1963. 1963 p.

[21] Chen F, Huss RS, Malone MF, Doherty MF. Simulation of kinetic effects in reactive distillation. Computers and Chemical Engineering. 200;**24**(11):2457–2472. doi:10.1016/S0098-1354(00)00609-8

[22] Yaws CL. Yaws' Handbook of Thermodynamics and Physical Properties of Chemical Compounds. New York: Knovel; 2003. 823 p.

[23] King CJ. Separation Processes. 2nd ed. New York: McGrawHill; 1980. 850 p.

Chapter 2

Short-Cut Methods for Multicomponent Batch Distillation

A. Narváez-García, J.C. Zavala-Loría,
A. Ruiz-Marín and Y. Canedo-López

Additional information is available at the end of the chapter

http://dx.doi.org/10.5772/66830

Abstract

This work present the batch distillation shortcut methods developed from the Fenske-Underwood-Gilliland (FUG) method of continuous distillation. In addition, a comparison between the results of shortcut method developed by Diwekar and Narváez-García et al. is done. The results of both methods are compared for validation with the results using the rigorous method presented by Domenech and Enjalbert. The results indicate no significant differences in both shortcut methods being the maximum deviation found between the two shortcut methods is less than 3%. Then, the use of any of both methods is in the ease or difficulty of the solution each of the model equations. Both methods were implemented using Fortran programming. Multdistbatch© software was used for the rigorous method.

Keywords: shortcut method, FUG, batch distillation

1. Introduction

Batch distillation is a process widely used for separation of small quantities of chemical compound of the one mixture as the recovery of small quantities of hazardous materials in waste streams, recovery of solvents, as well as, for pharmaceutical and biotechnological products with high added value, among others. Therefore, the development of mathematical models for the prediction of a process has a high interest in recent times [1–4].

Batch distillation is a flexible process because one equipment can obtain the separation of all the components of the mixture, while the continuous process [5] requires a number of columns distillation equal to the number of components minus one ($n - 1$). Another advantage of batch distillation process is the use of the same equipment for the mixture separation with different compositions or different mixtures [6].

On the other hand, the disadvantage of the batch distillation with respect to continuous distillation is that only small amounts of products can be obtained of the mixture. Another disadvantage is the production of waste unwanted for each cuts, however, these residual cuts can be separated into the same column [7].

A batch distillation column can be operated using any of the following policies [1]:

(1) Constant reflux.

(2) Variable reflux.

(3) Optimal reflux.

(4) Reflux profile.

The process behavior can be predicted by developing mathematical models based on mass and energy balances. The mathematical models obtained can be classified as [1, 8, 9]:

(1) Simplified (shortcut method).

(2) Semirigorous.

(3) Rigorous.

(4) Order reduction.

Currently the rigorous models have an area of great interest and these require especially the use of computers with high accuracy and processing capabilities; however, simplified methods can be applied with the use of the equipment such as tablets, smart phones, and/or laptops with smaller capacity of data processing, which makes possible the search for predicting the behavior of the process [2, 10]. In addition, the use of this kind of methods is a tool for obtaining initial data for the mathematical optimization.

Unlike the rigorous methods that considered the dynamics of the complete column, the shortcut methods are mathematical models that predict the behavior of the process considering the least amount of equations, usually making an overall material balance and partial balances considering a component any "*i*." The main limiting factor of these shortcut methods is to find a functional relationship between the concentrations of the dome and the bottom.

The shortcut methods are justified because these require a minor calculus time and memory requirements, as well as, an acceptable accuracy in the results obtained with respect to the rigorous method. These are an appropriate tool to obtain initial values for the mathematical optimization of some process, when the complexity of the methods required data very close to the solution.

The shortcut method is also used for the columns design and obtaining of limit conditions as minimum reflux ratio, R_{min} and minimum number of stage N_{min}. On the other hand, the shortcut methods are very simple to apply and to program, therefore, are useful in the teaching-learning process.

The two most important shortcut methods reported in the literature made use of the Fenske-Underwood-Gilliland (*FUG*) method developed for continuous distillation, but considering that the feed changes at every instant; that is, the bottom product in the current time is the feed for the next time (step).

The first of the shortcut methods was developed by Diwekar [11] and reported in the literature by Diwekar and Madhavan [12]. This method was developed considering the policies of constant and variable reflux. This method used the Hengstebeck-Geddes equation. This method also performs the comparison between the values of the minimum reflux ratio of Underwood and minimum reflux ratio of Gilliland, which increases the computational time because it uses an additional iterative process.

The second method was reported by Sundaram and Evans [13] and only considered the constant reflux policy and the Fenske Equation. The model obtains a solution in two parts; an outer loop that solves material balances and internal one that solves the functional relationship between the compositions of the dome and the bottom using the *FUG* method. The mathematical model developed initially considered:

(1) Constant relative volatilities.

(2) Constant molar flow.

(3) Negligible vapor and liquid accumulation in trays and the condenser.

Based on the work of Sundaram and Evans [13], Narváez-García et al. [10] developed a mathematical model for batch distillation process using a variable reflux policy.

The present studies show the most important shortcut methods used to predict the behavior of the batch distillation process.

2. Important definitions

For the use of the Underwood equations, this work considered separations Class I and Class II. In according to Shiras et al. [14] defined Class I and Class II as follows:

> **Class I**: "Separations such that, with infinite plates, all components of the feed are present in both the top product and bottom product."
>
> **Class II**: "Separations such that, with infinite plates, some of the components are completely in the top product or completely in the bottom product."

Similarly, an important concept in model developments is the key component light (*lk*) and heavy key component (*hk*) defined as follows:

> **Light key component (*lk*)**: Light component that is present in the residue in important amounts.
>
> **Heavy key component (*hk*)**: Heavy component that is present in the distillate in important amounts.

3. Reflux policies

When a fraction of the product obtained is fed back into the process and this can be done on four operations of the process: (1) constant reflux, (2) variable reflux, (3) optimum reflux, and (4) profile of reflux. In either case, the reflux ratio (R) is defined as

$$R = \frac{L}{\frac{dD}{dt}} = \frac{L}{D(t)} = \frac{L}{D} \tag{1}$$

where L is the reflux in the dome and D the product flow.

For some type of reflux used, it should be considered if there is accumulation of liquid and vapor in each of the trays as well as in the reflux tank. Another aspect that should be considered is where the initial feed is introduced because when it is performed from the reflux tank, the accumulation in each of the stages is equal to feed initial concentration, if conversely, the feed is introduced in the reboiler and column is operated without reflux, the concentration of each of the trays is equal to the concentration of the vapor phase and will be in equilibrium with the feed [15].

Constant reflux: In constant reflux policy, the product concentration varies with time because new feed input does not exist, so that the initial mole fractions of the more volatile are depleted and the molar fraction of the final distillate is an average.

Variable reflux: The batch distillation process with a variable reflux policy is used when it is desired to obtain a constant product concentration. In others words, reflux ratio is modified such that at each instant the same concentration of distillate is obtained.

Optimal reflux: For optimum reflux policy, the process used an objective function directly related to a control variable, which usually is the reflux ratio. This function is solved by applying mathematical methods such as dynamic programming, variation calculation, pontryagin maximum principle or nonlinear programming (NLP), among others. In general, the process is considered as an optimal control problem and the most common cases studied in the literature are [16]: (a) maximum distillate problem, (b) minimum time problem, (c) maximum profit problem, (d) minimum energy problem, and (e) maximum thermodynamic efficiency problem.

Reflux profile: For this case, a combination of constant reflux and variable reflux are used for obtaining a given concentration of the desired product in a time given. This operation policy is a derivation of the optimization process.

4. Materials and methods

In this work is considered a batch distillation column with the following characteristics:

- Adiabatic column.
- Theoretical trays.

- Partial reboiler.
- Total condenser.
- Constant pressure.
- Constant relative volatility.
- Negligible accumulation (holdup) of liquid and vapor.
- Constant molar flows through the column.

The mathematical model of the column is obtained by performing a total mass balance and partial mass balances to component "*i*." The Fenske-Underwood-Gilliland method is used to find the functional relationship between the compositions of the bottom and the dome of the column.

Although it presents the development of the model, considering the policies of operation of constant and variable reflux, these are presented in only four cases of study for the shortcut method to reflux variable considering the contribution to the state of the art of the authors.

In each case, it is considered that the mixture is fed to the boiling temperature. The error tolerance is 10^{-4}, the integration step is $\Delta t = 10^{-1}$ h and the time of production is required to deplete the lighter component. It has been considered a feed of 200 kmol and a vapor flow of 110 kmol h^{-1}. For these cases, it is considered that the relative volatility is constant throughout the process.

The value of the vapor flow was established so that it allows to deplete the most volatile component in a small operation time. For the ternary and quaternary mixtures only first cut is considered. For validation, the results of both methods, Diwekar [11] and Narváez-García et al. [10], are compared with the results using the rigorous method presented by Domenech and Enjalbert [17]. This model is used because it is considered as a low holdup. To solve each one of the cases was made a program in Fortran language.

5. Simplified mathematical models

The complete mathematical model of a batch distillation column considering the dynamics of the process consists of a system of differential equations and algebra (DAEs) added by equations that allow the calculation of the thermodynamic properties and hydraulic conditions of the column. The solution of the system can be very complex depending on the state equations used to predict the behavior of the gas phase (Soave, Redlich-Kwong Peng-Robinson, etc.) or the solution of the models used to predict the liquid phase behavior (Wilson, NRTL, UNIQUAC, UNIFAC, etc.).

According to Diwekar [9], the number of equations in a rigorous mathematical model of a batch distillation, which considers a mixture of components nc and N trays is equal to $(N + 2)(2nc + 1)$, where $N + 2$ considers the reboiler and condenser-tank reflux. Further, the number of equations consider the total restrictions in each of the stages ($\sum x = \sum y = 1$), the expression of reflux ratio ($R = L / D$), and liquid (L) and vapor (V) flows calculations along the column. The number of equations increases if the calculation of other variables of interest such as the column hydraulic or thermodynamic efficiency is considered.

The solution of this equation system is complex and requires intensive use of computers with adequate processing capacity, which affects costs in the area of process simulation. Therefore, it is needed to consider some simplifications to the mathematical model to reduce the data processing time.

Reductions to the batch distillation mathematical model are possible if it is considered that the process is continuous, with a feed that changes in every moment as shown in **Figure 1** [1, 10], which allows to use equations Fenske [18], Underwood [19], and Gilliland [20] of continuous distillation (*FUG* method).

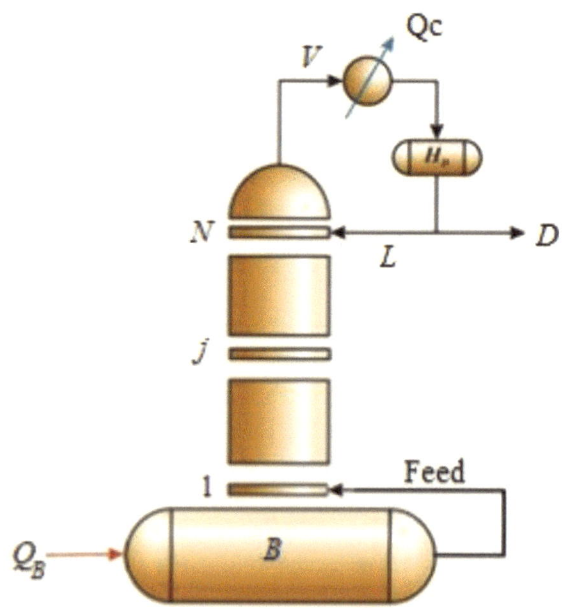

Figure 1. Scheme of a batch distillation column for the shortcut method.

Gilliland correlation can be replaced by the correlation Eduljee [21] because the mathematical expression is simpler for numerical works. The shortcut method considers:

(1) Constant molar flow along the column.

(2) Constant relative volatilities throughout the process.

(3) Negligible fluid and vapor accumulation within the column.

Constant molar flow is based on the assumption that the enthalpy of vaporization is the same for all components, which is correct if the mixture consists of very similar compounds.

The simplification is more restrictive from shortcut method to consider constant relative volatilities throughout the process. This consideration significantly reduces the number of calculations in the model, especially because the iterative processes of liquid-vapor equilibrium does not apply. When the relative volatility cannot be considered as a constant amount along the time or the column, polynomial expressions or Winn [22] equation can be used to estimate the changes; therefore, Diwekar [11] suggested that the relative volatilities can be used to calculate in every moment of the process using an average between the values of the bottom and the dome.

Finally, the vapor accumulation in a distillation column can be neglected because it is much less than the cumulative amount of liquid, and the accumulation of fluid in the column can be neglected, considering that this accumulation is less than the liquid accumulated in the reboiler. Under these circumstances, the two most important shortcut methods for batch distillation in the literature were reported by Diwekar [11], Sundaram and Evans [13], and Narvaez-García et al. [17].

5.1. Shortcut methods derived from FUG method

5.1.1. Shortcut method developed by Diwekar [11]

The first shortcut method for batch distillation presented here was developed by Diwekar [11]. This method considers a global balance in the column and their respective partial balances (component "*i*"); each of the equations used in the method are presented below:

Global balance:

$$\frac{dB}{dt} = -D; \qquad B_0 = F \tag{2}$$

where D is the distillate obtained by a mass balance in the dome of the column:

$$D = \frac{V}{R+1} \tag{3}$$

where R is the reflux ratio:

$$R = \frac{L}{D} \tag{4}$$

Partial balance with respect to a component (i):

$$\frac{d[B\,x_B^{(i)}]}{dt} = -D\,x_D^{(i)} \tag{5}$$

$$B\frac{d\,x_B^{(i)}}{dt} + x_B^{(i)}\frac{dB}{dt} = -D\,x_D^{(i)} \tag{6}$$

$$B\frac{d\,x_B^{(i)}}{dt} + x_B^{(i)}\frac{dB}{dt} = -D\,x_D^{(i)} = \left(\frac{dB}{dt}\right)x_D^{(i)} \tag{7}$$

and substituting in Eq. (2) is obtained:

$$B\frac{d\,x_B^{(i)}}{dt} - D\,x_B^{(i)} = -D\,x_D^{(i)} \tag{8}$$

substituting in Eq. (3) is obtained:

$$\frac{d\,x_B^{(i)}}{dt} = \frac{D}{B}\left[\,x_B^{(i)} - x_D^{(i)}\right]; \qquad x_{B_0}^{(i)} = x_{F_0}^{(i)} \tag{9}$$

$$\frac{d\,x_B^{(i)}}{dt} = \frac{V}{B(R+1)}\left[\,x_B^{(i)} - x_D^{(i)}\right]; \qquad x_{B_0}^{(i)} = x_{F_0}^{(i)} \tag{10}$$

A material balance is also performed for the reference component (k) and the ratio between the component (i) and the component (k).

$$\frac{d\,x_B^{(k)}}{dt} = \frac{D}{B}\left[\,x_B^{(k)} - x_D^{(k)}\right] = \frac{V}{B(R+1)}\left[\,x_B^{(k)} - x_D^{(k)}\right]; \qquad x_{B_0}^{(k)} = x_{F_0}^{(k)} \tag{11}$$

$$\frac{\frac{d\,x_B^{(i)}}{dt}}{\frac{d\,x_B^{(k)}}{dt}} = \frac{\frac{D}{B}\left[\,x_B^{(i)} - x_D^{(i)}\right]}{\frac{D}{B}\left[\,x_B^{(k)} - x_D^{(k)}\right]}; \qquad x_{B_0}^{(i)} = x_{F_0}^{(i)}; \quad i = 1, 2, \dots, n; i \neq k \tag{12}$$

$$\frac{d\,x_B^{(i)}}{d\,x_B^{(k)}} = \frac{\left[\,x_B^{(i)} - x_D^{(i)}\right]}{\left[\,x_B^{(k)} - x_D^{(k)}\right]}; \qquad x_{B_0}^{(i)} = x_{F_0}^{(i)}; \quad i = 1, 2, \dots, n; i \neq k \tag{13}$$

$$d\,x_B^{(i)} = \frac{\left[\,x_B^{(i)} - x_D^{(i)}\right]}{\left[\,x_B^{(k)} - x_D^{(k)}\right]}\,d\,x_B^{(k)}; \qquad x_{B_0}^{(i)} = x_{F_0}^{(i)}; \quad i = 1, 2, \dots, n; i \neq k \tag{14}$$

Eqs. (2) and (14) can be discretized if considered very small changes, then considering Eq. (2), is obtained:

$$\frac{\Delta B}{\Delta t} = -D \tag{15}$$

$$\Delta B = -D\Delta t \tag{16}$$

$$B_{new} = B_{old} - D\Delta t \tag{17}$$

Using Eq. (3) is obtained:

$$B_{new} = B_{old} - \left(\frac{V}{R+1}\right)\Delta t \tag{18}$$

Considering Eq. (14) is obtained:

$$\Delta\,x_B^{(i)} = \frac{\left[\,x_B^{(i)} - x_D^{(i)}\right]_{\text{new}}}{\left[\,x_B^{(k)} - x_D^{(k)}\right]_{\text{old}}}\,\Delta\,x_B^{(k)}; \qquad x_{B_0}^{(i)} = x_{F_0}^{(i)}; \quad i = 1, 2, \dots, n; i \neq k \tag{19}$$

$$x_{B,new}^{(i)} = x_{B,old}^{(i)} + \frac{\left[\,x_B^{(i)} - x_D^{(i)}\right]_{old}}{\left[\,x_B^{(k)} - x_D^{(k)}\right]_{old}}\,\Delta\,x_B^{(k)}; \qquad x_{B_0}^{(i)} = x_{F_0}^{(i)}; \quad i = 1, 2, \dots, n; i \neq k \tag{20}$$

The functional relationship between the concentrations of the dome and the bottom for this method is obtained using the equation Hengestebeck-Geddes:

$$x_D^{(i)} = \left(\frac{\alpha_i}{\alpha_k}\right)^{C_1}\left[\frac{x_D^{(k)}}{x_B^{(k)}}\,x_B^{(i)}\right]; \quad i = 1, 2, \dots, n; i \neq k \tag{21}$$

being α the relative volatility and C_1 a constant equal to the minimum number of trays of Fenske equation. It is also necessary to use the equations of Underwood and Gilliland or Eduljee:

$$\sum_{i=1}^{nc}\frac{\alpha_i\,x_B^{(i)}}{\alpha_i - \theta} = 0 \tag{22}$$

$$R_{\min} = -1 + \sum_{i=1}^{nc}\frac{\alpha_i\,x_D^{(i)}}{\alpha_i - \theta} \tag{23}$$

$$Y = 1 - \exp\left[\frac{(1 + 54.4X)(X - 1)}{(11 + 117.2X)\sqrt{X}}\right] \tag{24}$$

The Eduljee correlation is

$$Y = 0.75(1 - X^{0.5668}) \tag{25}$$

being, in both cases:

$$X = \frac{R - R_{\min}}{R + 1} \tag{26}$$

$$Y = \frac{N - N_{\min}}{N + 1} \tag{27}$$

In the case of a constant reflux policy, Diwekar [11] uses a function that relates the minimum reflux ratio obtained with Underwood equation (R_{minU}) and the minimum reflux ratio obtained with Gilliland Equation (R_{minG}) can be expressed as

$$f_c = \frac{R_{\mathrm{minU}}}{R} - \frac{R_{\mathrm{minG}}}{R} = 0 \tag{28}$$

In the variable reflux case, Diwekar [11] uses a function obtained from the equation Hengestebeck-Geddes (Eq. (21)), considering the sum of all components, and applying the sum in both members of this equation is obtained:

$$\sum_{i=1}^{nc} x_D^{(i)} = \sum_{i=1}^{nc} \left(\frac{\alpha_i}{\alpha_k}\right)^{C_1} \left[\frac{x_D^{(k)}}{x_B^{(k)}} x_B^{(i)}\right] = 1 \tag{29}$$

$$\frac{x_B^{(k)}}{x_D^{(k)}} = \sum_{i=1}^{nc} \left(\frac{\alpha_i}{\alpha_k}\right)^{C_1} x_B^{(i)} \tag{30}$$

Therefore, the function is

$$f_v = \left\{ \sum_{i=1}^{nc} \left[\left(\frac{\alpha_i}{\alpha_k}\right)^{C_1} x_B^{(i)} \right] \right\} - \frac{x_B^{(k)}}{x_D^{(k)}} \tag{31}$$

In this method, it is necessary to consider a reference component, hence, using the equation Hengestebeck-Geddes (Eq. (21)) and separating the concentration of the reference component (k):

$$x_D^{(k)} = \frac{x_B^{(k)}}{\sum_{i=1}^{nc} \left[\left(\frac{\alpha_i}{\alpha_k}\right)^{C_1} x_B^{(i)} \right]} \tag{32}$$

With the developed mathematical models, it is necessary to consider finding the solution; therefore, the solution algorithm is shown below.

5.1.1.1. Diwekar solution algorithm

The shortcut method developed by Diwekar [11] is setup by the system of Eqs. (1)–(32). To solve this system of equations, it is necessary to follow the next steps:

Constant reflux policy.

(1) Steady-state concentrations are calculated considering that $C_1 = N$.

  (a) The concentration of component reference (k) in the dome (Eq. (32)) is calculated.

  (b) Other concentrations are calculated using the Hengestebeck-Geddes equation (Eq. (22)).

  (c) Increase the time (Δt).

  (d) New concentrations (Eq. (20)) and the remaining amount in the reboiler (Eq. (18)) are calculated.

(2) To propose an initial value of C1, which will be adjusted by an iterative process.

(3) To calculate the reference component concentration in the dome (Eq. (32)).

(4) Other concentrations are calculated using Equation Hengestebeck-Geddes (Eq. (21)).

(5) To solve Underwood equations (22) and (23).

(6) To solve Gilliland equation.

   (a) First Eq. (24).

   (b) Second Eq. (26).

   (c) Finally, Eq. (27).

(7) Verify that the obtained value by Underwood equations is the same to that obtained by Equation Gilliland.

   (a) To use Eq. (31).

   (b) If this is not true it is necessary to change the value of C_1 with some iterative process as the Newton-Raphson method.

   (c) The process is repeated from step 2 to achieve convergence.

   (d) If this is true go to step 8.

(8) Increase the time (Δt).

(9) Calculate new concentrations (Eq. (20)) and the remaining amount in the reboiler (Eq. (18)).

(10) The process is repeated until the desired time production.

Variable reflux policy.

(1) Steady-state concentrations are calculated considering that $C_1 = N$.

   (a) Consider constant the concentration of reference component (k) in the dome.

   (b) Other concentrations are calculated using the equation Hengestebeck-Geddes equations (Eq. (21)).

   (c) Increase the time (Δt).

   (d) New concentrations (Eq. (20)) and the remaining amount in the reboiler (Eq. (18)) are calculated.

(2) To propose an initial value of C_1, which will be adjusted by an iterative process.

(3) Verify that Eq. (31) is zero.

   (a) If this is not true it is necessary to change the value of C_1 with some iterative process as the Newton-Raphson method until converge.

(4) Other concentrations are calculated using Equation Hengestebeck-Geddes (Eq. (21)).

(5) To solve Underwood equations (22) and (23).

(6) To solve Gilliland equation.

(a) First Eq. (27).

(b) Second Eq. (24).

(7) Calculate the value of the reflux ratio R with Eq. (26).

(8) Increase the time (Δt).

(9) Calculate new concentrations (Eq. (20)) and the remaining amount in the reboiler (Eq. (18)).

(10) The process is repeated until the desired time production.

5.1.2. *Shortcut method developed by Sundaram and Evans [13] using a constant reflux policy*

In the method of Sundaram and Evans [13], the total material balance (Eq. (2)) and partial (Eq. (4)) are similar to the method of Diwekar [11], and Eqs. (18) and (20) are the same; however, Eq. (20) may be a function of the remaining liquid in the bottom; therefore, Eq. (6) calculates the change in the mole fractions in the bottom, then:

$$d\,x_B^{(i)} = [x_D^{(i)} - x_B^{(i)}]\frac{dB}{B} \tag{33}$$

Considering very small changes in the above equation; therefore, it is obtained:

$$\Delta\,x_B^{(i)} = [x_D^{(i)} - x_B^{(i)}]_{\text{old}}\frac{\Delta B}{B_{\text{old}}} \tag{34}$$

$$x_{B,new}^{(i)} = x_{B,old}^{(i)} + [x_D^{(i)} - x_B^{(i)}]_{old}\frac{(B_{new} - B_{old})}{B_{old}}; \quad x_{B_0}^{(i)} = x_{F_0}^{(i)}\,;\, i = 1, 2, \ldots, n \tag{35}$$

Using the reference component (k) in the partial balance, instead of component (i), it is obtained:

$$d\,x_B^{(k)} = [x_D^{(k)} - x_B^{(k)}]\frac{dB}{B} \tag{36}$$

and considering very small changes in Eq. (24) can be obtained:

$$\frac{(B_{new} - B_{old})}{B_{old}} = \frac{\Delta\,x_B^{(k)}}{[x_D^{(k)} - x_B^{(k)}]_{old}} \tag{37}$$

Substituting Eq. (37) into Eq. (35), Eq. (20) is obtained. Eq. (28) is easily solved; however, Eq. (30) and Eq. (35) are much more complex because they require the functional relationship between the concentrations of the bottom and the dome.

The functional relationship between the concentrations of the dome and the bottom is calculated using the Fenske equation considering the minimum number of separation stages (N_{min}) with the mole fractions of the dome (x_D) and bottom (x_B):

$$N_{\text{min}} = \frac{\ln\left[\frac{x_D^{(i)}\, x_B^{(k)}}{x_D^{(k)}\, x_B^{(i)}}\right]}{\ln(\alpha_i)} \tag{38}$$

The Underwood equation for Class I mixtures and the Gilliland or Eduljee equations are used. In the original work of Sundaram and Evans [13] the equations for mixtures Class II were not considered.

When the mixture is classified as a multicomponent system Class I, the Underwood equation that relates the R_{min} it can be expressed as [22]:

$$R_{min} = \frac{\left[\frac{x_D^{(lk)}}{x_B^{(lk)}}\right] - \alpha_{lk,lk}\left[\frac{x_D^{(hk)}}{x_B^{(hk)}}\right]}{\alpha_{lk,hk} - 1} \qquad (39)$$

In this method, it is also necessary to consider the composition of a component (k) of reference; then, using the Fenske equation, the composition of the reference component is isolated. Therefore, it is necessary to considerer the sum of all components. From Eq. (35), the following is obtained:

$$x_D^{(i)} = x_B^{(i)}\left[\frac{x_D^{(k)}}{x_B^{(k)}}\right]\alpha_i^{N_{min}}; \quad i = 1, 2, \ldots, n; i \neq k \qquad (40)$$

$$\sum_{i=1}^{nc} x_D^{(i)} = \sum_{i=1}^{nc} x_B^{(i)}\left[\frac{x_D^{(k)}}{x_B^{(k)}}\right]\alpha_i^{N_{min}} = 1 \qquad (41)$$

$$x_D^{(k)} = \frac{x_B^{(k)}}{\sum_{i=1}^{nc}(\alpha_i^{N_{min}} x_B^{(i)})} \qquad (42)$$

With all the mathematical equations of the shortcut method, the next step is to provide an appropriate methodology for the solution.

5.1.2.1. Sundaram and Evans [13] solution algorithm

The mathematical model of Sundaram and Evans is formed by the system of Eqs. (35), (37)–(42). To solve this system, follow the next steps:

(1) Steady-state concentrations are calculated considering that N_{min}=N.

  (a) The concentration of reference component (k) in the dome (Eq. (42)) is calculated.

  (b) Other concentrations are calculated using Fenske Eq. (40).

  (c) Increase the time (Δt).

  (d) New concentrations (Eq. (35)) and the remaining amount in the reboiler (Eq. (37)) are calculated.

(2) The R_{min} and N_{min} are calculated.

  (a) For R_{min} Eq. (39) is used.

  (b) For N_{min} Eq. (38) is used.

  (c) Eqs. (38) and (39) are solved using an iterative process.

(3) Calculate the reference component concentration in the dome (Eq. (42)).

(4) The other concentrations are calculated using the Fenske Eq. (40).

(5) Back to step 2 until achieve the desired production time.

5.1.3. *Short method developed by Narváez-García et al. [10] using a variable reflux policy*

This proposal is based on the concepts of Sundaram and Evans [13]. It is initiated by calculating the reflux ratio required to obtain the desired product; therefore, using Eq. (26) and solving it, the following is obtained:

$$R = \frac{X + R_{\min}}{1 - X} \tag{43}$$

Eq. (43) requires the calculation of $R_{\min}$, which you can get from one of the equations of Underwood (Eq. (23)) as shown below:

$$R_{\min} = -1 + \sum_{i=1}^{nc} \frac{\alpha_i x_D^{(i)}}{\alpha_i - \theta} \tag{44}$$

Eq. (23) requires the calculation of θ, so Eq. (22) is used:

$$\sum_{i=1}^{nc} \frac{\alpha_i x_B^{(i)}}{\alpha_i - \theta} = 0 \tag{45}$$

Also, it is necessary to obtain the concentrations in the dome, therefore using the Fenske Equation (38) for the component (i):

$$x_D^{(i)} = x_B^{(i)} \left[\frac{x_D^{(k)}}{x_B^{(k)}} \right] \alpha_i^{\ N_{\min}}; \quad i = 1, 2, \ldots, n; i \neq k \tag{46}$$

Applying a sum process considering all the components of the mixture and calculating the composition of the reference component (k) and is obtained:

$$x_D^{(k)} = \frac{x_B^{(k)}}{\sum_{i=1}^{nc} \left(\alpha_i^{\ N_{\min}} x_B^{(i)} \right)} \tag{47}$$

Eqs. (40) and (42) require the calculation of the $N_{\min}$. Eq. (23) also requires the calculation of X; therefore, using Eqs. (25) and (27), the following is obtained;

$$X = \left[1 - \left(\tfrac{4}{3} \right) \left(\frac{N - N_{\min}}{N + 1} \right) \right]^{1.7643} \tag{48}$$

Eq. (48) requires the $N_{\min}$ value, therefore, it is necessary to obtain $N_{\min}$ for the shortcut method. In this sense the Fenske equation allows to calculate the minimum number of trays when the light key component (lk) and the heavy key component ($hk = k$) are considered, then:

$$N_{\min} = \frac{\log \left\{ \left[\frac{x_D^{(lk)}}{x_D^{(k)}} \right] \left[\frac{x_B^{(k)}}{x_B^{(lk)}} \right] \right\}}{\log(\alpha_{lk,k})} \tag{49}$$

where:

$$x_D^{(lk)} = \left[\frac{x_D^{(k)}}{x_B^{(k)}} \right] x_B^{(lk)} \alpha_{lk,k}^{N_{\min}} \tag{50}$$

To change Eq. (49) in function of known values, $x_D^{(k)}$ should be replaced by an $x_D^{(lk)}$ expression, therefore, knowing that $\sum_{i=1}^{nc} x_D^{(i)} = 1$, then:

$$\sum_{i=1}^{nc} x_D^{(i)} - x_D^{(lk)} = 1 - x_D^{(lk)} \tag{51}$$

Substituting Eq. (40) into Eq. (51), the following is obtained:

$$\sum_{i=1}^{nc}\left\{\left[\frac{x_D^{(k)}}{x_B^{(k)}}\right](\alpha_{i,k})^{N_{\min}}x_B^{(i)}\right\}-x_D^{(lk)} = 1-x_D^{(lk)} \tag{52}$$

Substituting Eq. (50) in Eq. (52), the following is obtained:

$$\sum_{i=1}^{nc}\left\{\left[\frac{x_D^{(k)}}{x_B^{(k)}}\right](\alpha_{i,k})^{N_{\min}}x_B^{(i)}\right\}-\left[\frac{x_D^{(k)}}{x_B^{(k)}}\right]x_B^{(lk)}\,\alpha_{LK,k}^{N_{\min}} = 1-x_D^{(lk)} \tag{53}$$

$$x_D^{(k)} = \frac{[1-x_D^{(lk)}]x_B^{(k)}}{\sum_{i=1}^{nc}\alpha_{i,k}^{N_{\min}}x_B^{(i)}-x_B^{(lk)}\,\alpha_{lk,k}^{N_{\min}}} = \frac{[1-x_D^{(lk)}]x_B^{(k)}}{\sum_{i=2}^{nc}\alpha_{i,k}^{N_{\min}}x_B^{(i)}} \tag{54}$$

and, Eq. (49) becomes:

$$N_{\min} = \frac{\log\left\{\left[\frac{x_D^{(lk)}\sum_{i=2}^{nc}\alpha_{i,k}^{N_{\min}}x_B^{(i)}}{[1-x_D^{(lk)}]x_B^{(lk)}}\right]\right\}}{\log(\alpha_{lk,k})} \tag{55}$$

Eq. (55) requires an iterative process for the solution of $N_{\min}$; therefore, using the Newton-Raphson method, the following is obtained:

$$1-x_D^{(lk)} = \left[\frac{x_D^{(lk)}}{x_B^{(lk)}}\right]\sum_{i=2}^{nc}\left(\frac{\alpha_{i,k}}{\alpha_{lk,k}}\right)^{N_{\min}}x_B^{(i)} \tag{56}$$

$$f(N_{\min})=\left\{\left[\frac{x_D^{(lk)}}{x_B^{(lk)}}\right]\sum_{i=2}^{nc}\left[x_B^{(i)}\left(\frac{\alpha_{i,k}}{\alpha_{lk,k}}\right)^{N_{\min}}\right]\right\}-[1-x_D^{(lk)}]=0 \tag{57}$$

$$f'(N_{\min})=\left[\frac{x_D^{(lk)}}{x_B^{(lk)}}\right]\left\{\left[\sum_{i=2}^{nc}x_B^{(i)}\left(\frac{\alpha_{i,k}}{\alpha_{lk,k}}\right)^{N_{\min}}\right]\times\ln\left(\frac{\alpha_{i,k}}{\alpha_{lk,k}}\right)\right\} \tag{58}$$

$$(N_{\min})_{\text{actual}} = (N_{\min})_{\text{anterior}}-\frac{f(N_{\min})_{\text{anterior}}}{f'(N_{\min})_{\text{anterior}}} \tag{59}$$

With the $N_{\min}$ value, the reflux ratio and other values relating to the variable $N_{\min}$ can be calculated. The proposed solution to the developed method is addressed in the following section.

It is notable that both the model of Diwekar [11] and this model started from the same material balances (global and partial), and in other words, both works are developed following the same method; however, the functional relationship between the concentrations of the dome and bottom is different equations. **Table 1** presents a comparison between the equations of the two models.

In fact the equations of Underwood and Gilliland are the same in each model, and the difference is the way of how the values of the $N_{\min}$ are obtained. Narváez-García et al. [10] used the Fenske equation, while Diwekar [11] used the equation of Hengestebeck-Geddes.

While calculation times are similar in both models, the Narváez-García et al. model has an advantage over the Diwekar model when the separation of mixtures Class I is performed due to the use of a simplified Underwood equation (Eq. (39)). This does not happen with the model of Diwekar because the original equations of Underwood are always considered.

Diwekar [11]	Narváez-García et al. [10]
$f_v = \left\{\sum_{i=1}^{nc}\left[x_B^{(i)}\left(\frac{\alpha_i}{\alpha_k}\right)^{C_1}\right]\right\} - \frac{x_B^{(k)}}{x_D^{(k)}} = 0$	$f(N_{min}) = \left\{\left[\frac{x_D^{(lk)}}{x_B^{(lk)}}\right]\sum_{i=2}^{nc}\left[x_B^{(i)}\left(\frac{\alpha_{i,k}}{\alpha_{lk,k}}\right)^{N_{min}}\right]\right\} - [1-x_D^{(lk)}] = 0$
$x_D^{(i)} = x_B^{(i)}\left[\frac{x_D^{(k)}}{x_B^{(k)}}\right]\left(\frac{\alpha_i}{\alpha_k}\right)^{C_1};$ $i = 1, 2, \dots, n;\ i \neq k$	$x_D^{(i)} = x_B^{(i)}\left[\frac{x_D^{(k)}}{x_B^{(k)}}\right]\alpha_i^{N_{min}};$ $i = 1, 2, \dots, n\ ;\ i \neq k$
$x_D^{(k)} = \frac{x_B^{(k)}}{\sum_{i=1}^{nc}(\alpha_i^{N_{min}} x_B^{(i)})}$	$x_D^{(k)} = \frac{[1-x_D^{(lk)}]x_B^{(k)}}{\sum_{i=2}^{nc}\alpha_{i,k}^{N_{min}} x_B^{(i)}}$
$B_{actual} = B_{anterior} - \left(\frac{V}{R+1}\right)\Delta t$	$B_{actual} = B_{anterior} + \frac{\Delta x_B^{(k)}}{[x_D^{(k)} - x_B^{(k)}]_{anterior}} B_{anterior}$
$x_{B,actual}^{(i)} = x_{B,anterior}^{(i)} + \frac{[x_B^{(i)} - x_D^{(i)}]_{anterior}}{[x_B^{(k)} - x_D^{(k)}]_{anterior}}\Delta x_B^{(k)};$	$x_{B,actual}^{(i)} = x_{B,anterior}^{(i)} + [x_D^{(i)} - x_B^{(i)}]_{anterior}\frac{(B_{actual} - B_{anterior})}{B_{anterior}};$
$x_{B_0}^{(i)} = x_{F_0}^{(i)}\ ;\ i = 1, 2, \dots, n\ ;\ i \neq k$	$x_{B_0}^{(i)} = x_{F_0}^{(i)}\ ;\ i = 1, 2, \dots, n$
$\sum_{i=1}^{n}\frac{\alpha_i x_B^{(i)}}{\alpha_i - \theta} = 0;\ R_{min} = -1 + \sum_{i=1}^{n}\frac{\alpha_i x_D^{(i)}}{\alpha_i - \theta}$; CLASS II	
Not considered	$R_{min} = \frac{\left[\frac{x_D^{(lk)}}{x_B^{(lk)}}\right] - \alpha_{lk,lk}\left[\frac{x_D^{(hk)}}{x_B^{(hk)}}\right]}{\alpha_{lk,hk} - 1}$; CLASS I
$Y = 1 - \exp\left[\frac{(1+54.4X)(X-1)}{(11+117.2X)\sqrt{X}}\right];\ X = \frac{R - R_{min}}{R+1};\ Y = \frac{N - N_{min}}{N+1}$	

Table 1. Comparing the Diwekar [11] and Narváez-García et al. [10] shortcut method using variable reflux policy.

5.1.3.1. Solution algorithm (Narváez-García et al. [1])

The mathematical model by Narváez-García et al. [10] is conformed for the system of Eqs. (22) and (23) or (39), (35), (37), (43), (48), (49), (54), (55), (57), (58), and (59). The main objective of this system of equations is to calculate the value of the reflux ratio and for this Eq. (40) is used.

Eq. (43) requires the value of X and R_{min}. The value of X is related to the minimum number of trays (N_{min}) through Eqs. (48) and (49); therefore, first N_{min} is calculated, starting with an assumed value and is iterated until it converges to the correct value of N_{min}.

The Newton-Raphson iterative method used Eqs. (57)–(59). These equations are only function of the dome and bottom concentrations of as well as of the relative volatilities.

The value obtained of X allows to find the value of R_{min}, which is solve using the Underwood equation (39). However, to get the value of R_{min} before it is necessary to calculate the mole fractions of the dome (x_D) using Eqs. (40) and (54). With the values of X and R_{min} will be calculated the reflux ratio R (Eq. (43)) and now it is possible to calculate the amount remaining in the reboiler using Eq. (37) and the bottom concentration using Eq. (35).

6. Cases of study

The mathematical models of the shortcut method presented in this work have been solved considering various mixtures: binary, ternary, and quaternary. Being the variable reflux policy more complicated than the constant reflux policy, only are presented cases considering the variable reflux policy. The input conditions to the process are shown in **Table 2**.

Case	Feed molar fraction				Relatives volatilities (α)				N^+	k	$x_D^{(k)}$
	1	2	3	4	1	2	3	4			
1	0.40	0.20	0.30	0.10	1.67	1.25	1.00	0.83	5	3	0.70
2	0.33	0.33	0.34	–	1.33	1.00	0.67	–	10	2	0.80
3	0.50	0.50	–	–	2.40	1.00	–	–	9	2	0.95

N^+ = Number of trays, k= Reference, Component = 1, 2, 3, 4.

Table 2. Input conditions for cases of study.

7. Results and discussion

The results of the cases of the study are shown below. Considering that, the mole fraction of the desired component is a constant amount, the profiles of the reflux ratios, the remaining amounts in the bottom and its concentrations are obtained. To validate the results of the reflux rate obtained by the shortcut methods, a comparison between the profile of the reflux ratio obtained and the profile obtained with a rigorous method was performed.

7.1. Case 1

Figures 2, **4**, and **5** show the results obtained with the short methods of Diwekar [11] and Narváez-García et al. [10]. **Figure 3** shows the result comparison between the shortcut method and the rigorous method.

The comparison of the results between the two shortcut methods (**Figures 2**, **4**, and **5**) allows to establish that there are no significant differences. The maximum deviation for reflux was 1.5%, the amount remaining in the reboiler was 0.55 %, and concentrations in the bottom was 2%.

http://dx.doi.org/10.5772/66830

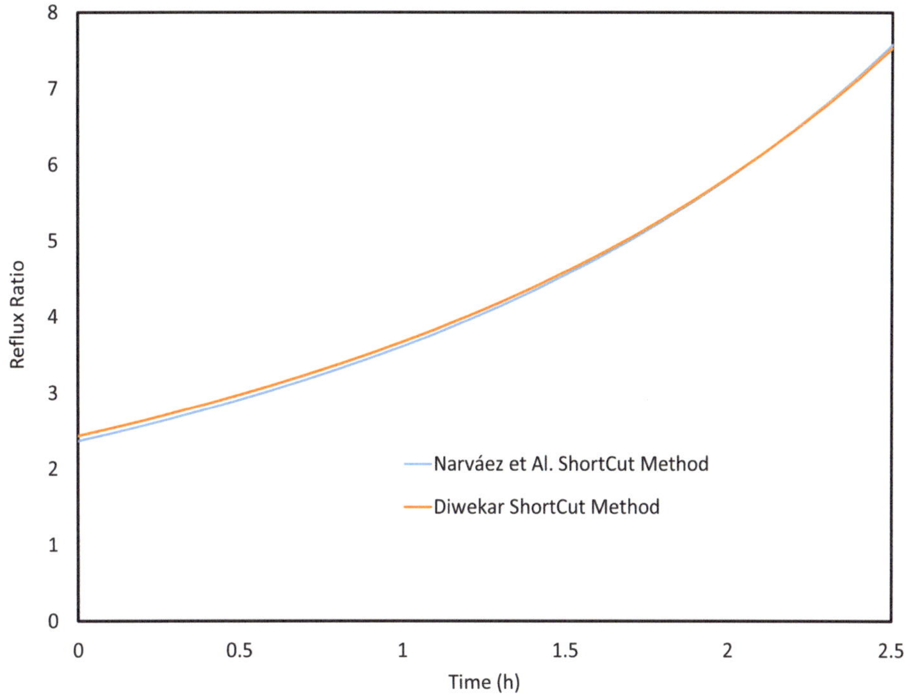

Figure 2. Reflux ratio profiles obtained with the shortcut methods.

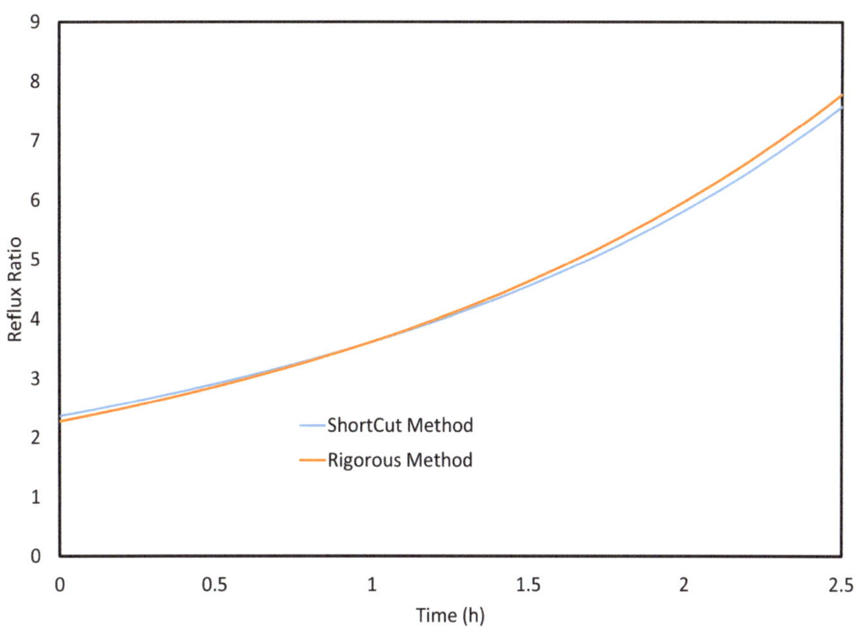

Figure 3. Comparison of profiles of reflux ratio using the shortcut method and a rigorous method.

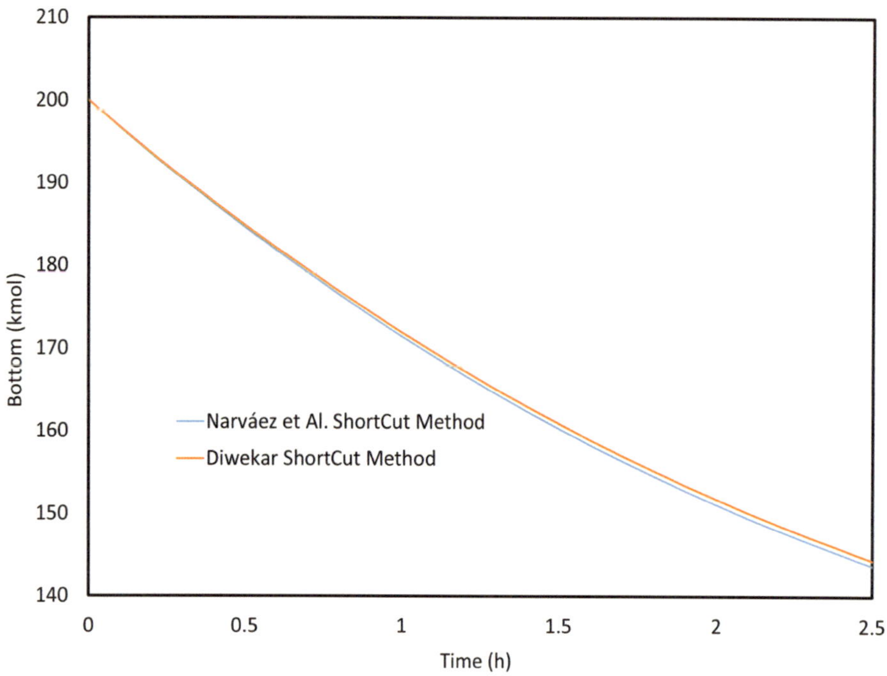

Figure 4. Remaining amount in the reboiler obtained by shortcut methods.

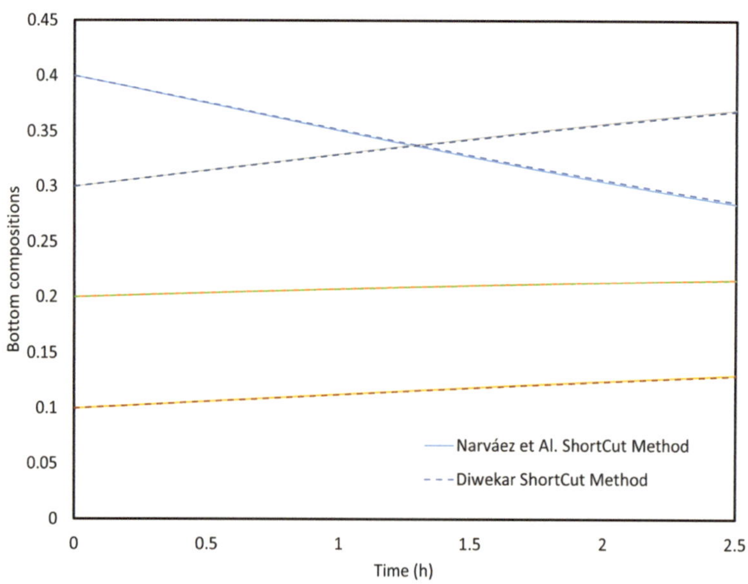

Figure 5. Concentration profiles in the reboiler obtained by shortcut methods.

As for the comparison between the shortcut method and the rigorous method (**Figure 3**), the deviations are within an acceptable range of 9.7% maximum considering the reflux ratio is calculated.

7.2. Case 2

Figures 6, **8**, and **9** show the results obtained with the short methods of Diwekar [11] and Narváez-García et al. [10]. **Figure 3** shows the results comparison between the shortcut method and the rigorous method.

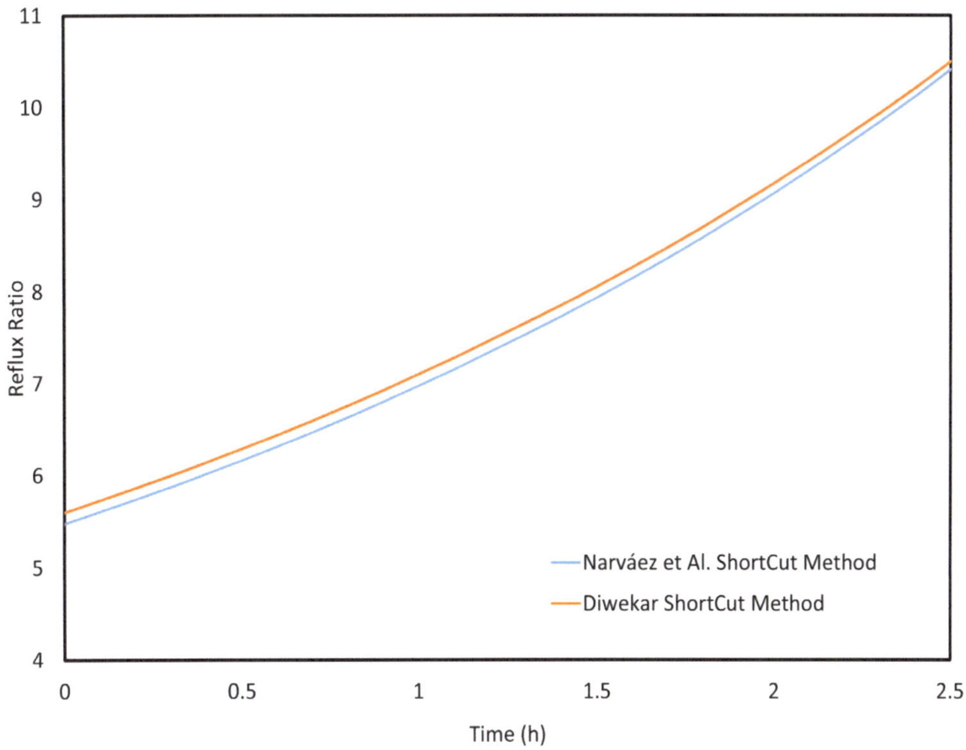

Figure 6. Reflux ratio profiles obtained by shortcut methods.

The results between both short methods (**Figures 6**, **8**, and **9**) allow to establish that there are no significant differences. The maximum deviation for calculated reflux ratio was 2.2%, for the amount remaining in the reboiler was 0.29%, and the deviation in the bottom concentrations was 0.67%. As for the comparison between the shortcut method and the rigorous method (**Figure 7**), the deviations are within an acceptable range of 9.7% maximum considering the reflux ratio is calculated.

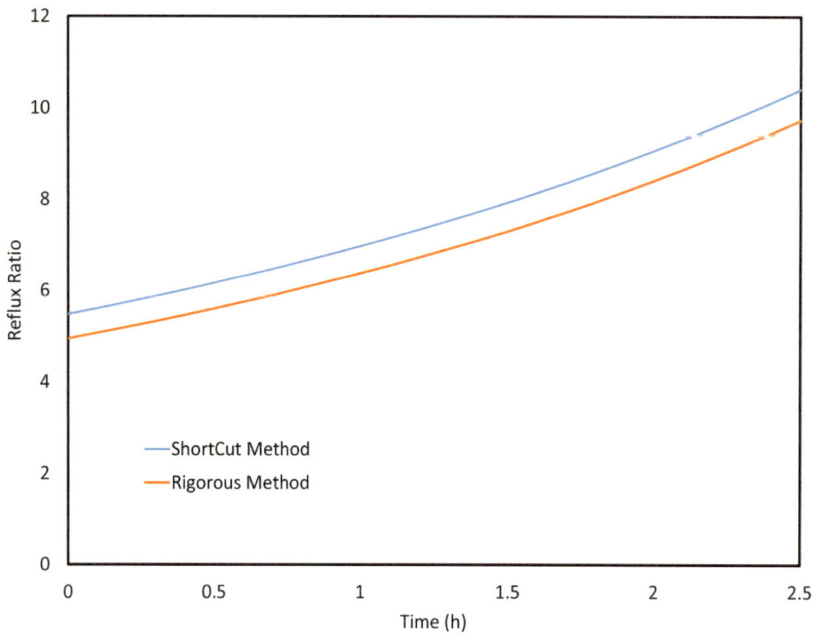

Figure 7. Comparison of profiles of reflux ratio obtained using the shortcut method and a rigorous method.

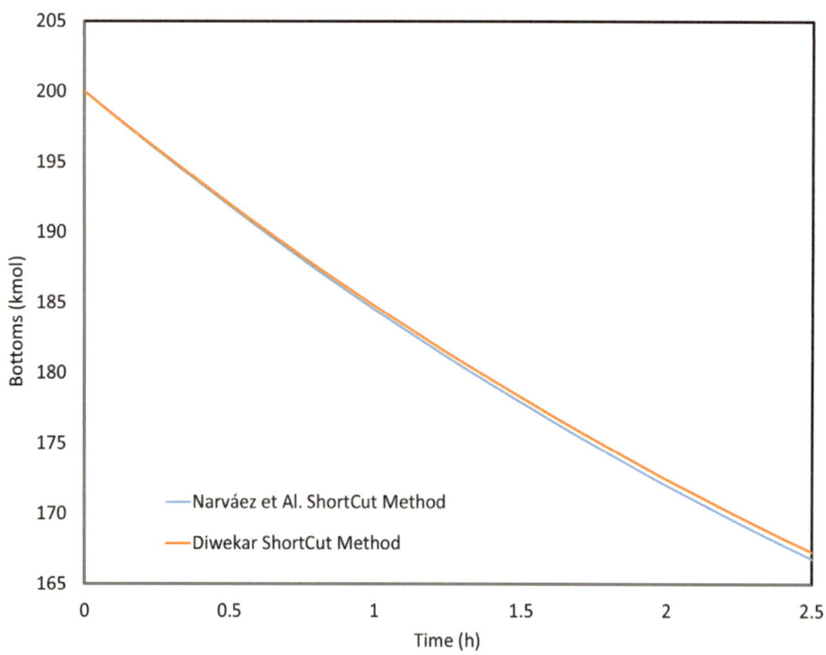

Figure 8. Remaining amount in the reboiler obtained by shortcut methods.

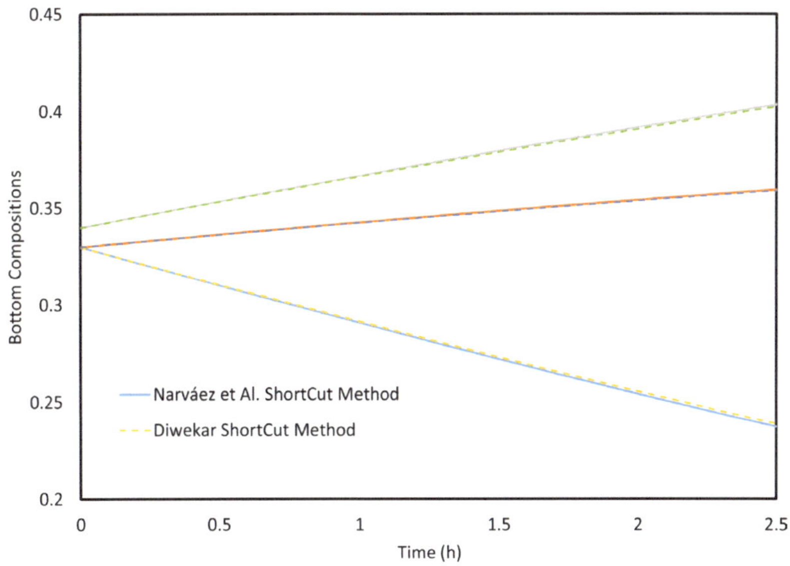

Figure 9. Concentration profiles in the reboiler obtained by shortcut methods.

7.3. Case 3

Figures 10, **12**, and **13** show the results obtained with the short methods of Diwekar [11] and Narváez-García et al. [10]. **Figure 11** shows the results comparison between the shortcut method and the rigorous method.

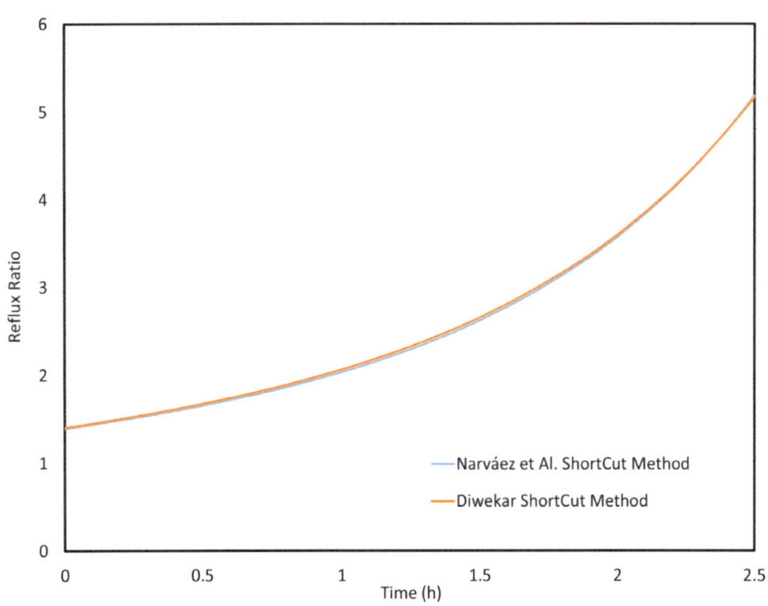

Figure 10. Reflux ratio profiles obtained by shortcut methods.

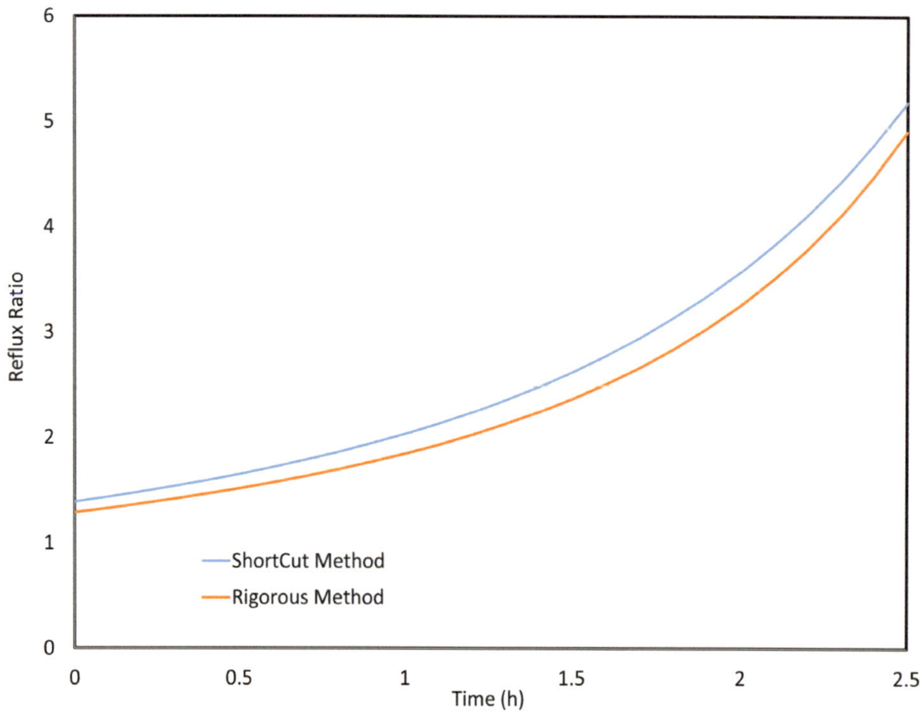

Figure 11. Comparison of profiles of reflux ratio obtained using the shortcut method and a rigorous method.

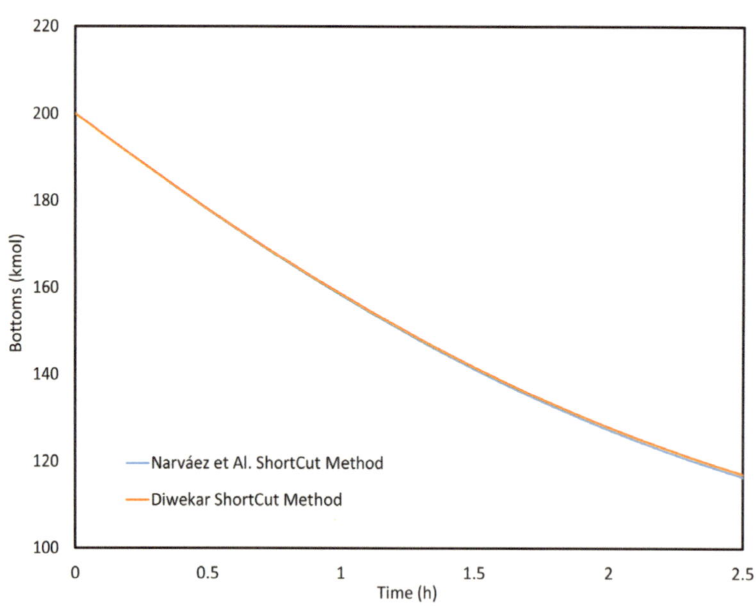

Figure 12. Remaining amount in the reboiler obtained by shortcut methods.

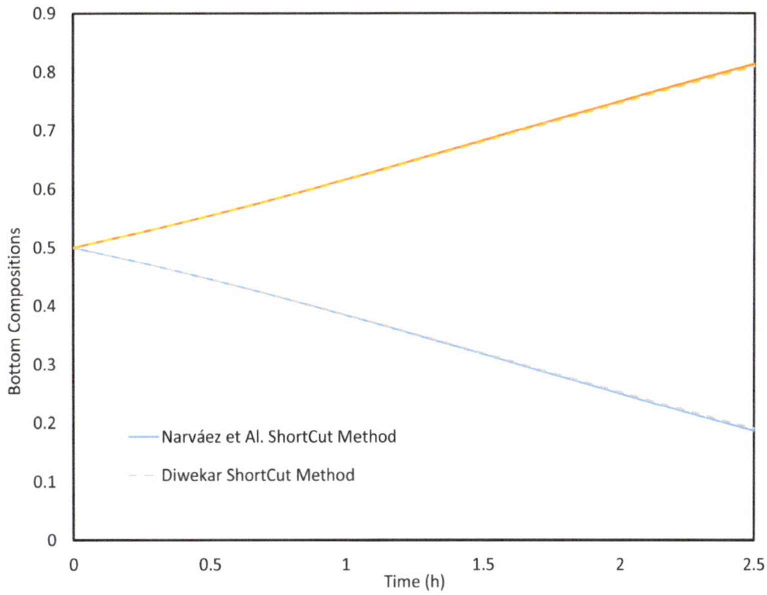

Figure 13. Concentration profiles in the reboiler obtained by shortcut methods.

The results between both short methods (**Figures 10**, **12**, and **13**) allow to establish that there are no significant differences. The maximum deviation for calculated reflux ratio was 2.7%, the amount remaining in the reboiler was 0.45%, and the deviation in the bottom concentrations was 0.63%. As for the comparison between the shortcut method and the rigorous method (**Figure 11**), the deviations are within an acceptable range of 3.8% maximum considering the reflux ratio is calculated.

In general, as shown in each of the figures, the maximum deviation found between the two shortcut methods considering a policy of variable reflux is less than 3% and, in this sense, the use of either of the two depends on the ease of application of the method.

In this case, the method developed by Narváez-García et al. [10] is better because it is adjusted to the original equations of the FUG method.

Similarly, to validate the shortcut methods considering a variable reflux policy, we had presented a comparison between the profiles of the reflux ratio which shows that you can have up to 9.7% difference between the results of the shortcut method and the rigorous method, of course, this difference is due to the simplifications of the short method, however, the difference falls within an acceptable range and this validated the shortcut methods presented. The maximum difference found between the concentrations of the bottom was less than 2%.

In all cases, the behavior of the profiles is adequate for the batch distillation process; in other words, greater process time is necessary for a greater reflux and the more volatile component is depleted.

8. Conclusions

In this chapter, we have presented the shortcut methods developed by Diwekar [11], Sundaram and Evans [13], and Narváez-García et al. [10]. Considering the complexity of the solution only, the shortcut method with a variable reflux policy is solved. The results were validated using a rigorous method. It is considered that the results of the shortcut methods are very close with respect to the rigorous method results.

Author details

A. Narváez-García[1], J.C. Zavala-Loría[2*], A. Ruiz-Marín[1] and Y. Canedo-López[1]

*Address all correspondence to: jose.zavala@unini.edu.mx

1 Faculty of Chemical Sciences and Petroleum, Autonomous University of Carmen, Carmen City, Campeche, Mexico

2 Iberoamerican International University, San Francisco de Campeche, Campeche, Mexico

References

[1] Narváez-García, A., Zavala-Loría,J.C., Rocha-Uribe, J.A. & Rubio-Atoche, C.M.; *Método Corto para la Destilación Discontinua Multicomponente Considerando una Política de Reflujo Variable;* Revista Mexicana de Ingeniería Química. **2013**, 12(3), pp. 621–637.

[2] Jain, S., Kim, J.K. & Smith, R.; *Operational Optimization of Batch Distillation Systems;* Industrial & Engineering Chemistry Research. **2012**, 51, pp. 5749–5761.

[3] Simon, L.L., Kence, H. & Hungerbuhler, K.; *Optimal Rectification Column, Reboiler Vessel, Connection Pipe Selection and Optimal Control of Batch Distillation Considering Hydraulic Limitations;* Chemical Engineering Processing. **2009**, 48, pp. 938–949.

[4] Zavala-Loría, J. C. & Coronado, C.; *Optimal Control Problem in Batch Distillation Using Thermodynamic Efficiency;* Industrial Engineering Chemistry. Research. **2008**, 47, pp. 2788–2793.

[5] Al-Tuwaim, M.S. & Luyben, W.L.; *Multicomponent Batch Distillation: 3. Shortcut Design of Batch Distillation Columns;* Industrial Engineering Chemistry and Research. **1991**, 30, pp. 507–516.

[6] Zavala Loría, J. C.; *Optimización del Proceso de Destilación Discontinua;* Tesis de Doctorado. Instituto Tecnológico de Celaya Gto, México. **2004**.

[7] Barolo, M. & Guarise, B.; *Batch Distillation of Multicomponent Systems with Constant Relative Volatilities;* Chemical Engineering. **1996**, 74 (8), pp. 863–871.

[8] Jain, S., Kim, J.K. & Smith, R.; *Process Synthesis of Batch Distillation Systems*; Industrial & Engineering Chemistry Research. **2013**, 52, pp. 8272–8288.

[9] Diwekar, U. M.; *Batch Distillation: Simulation, Optimal Design and Control*; 2da Edition, U.S.A.: Editorial CRC Press. **2012**.

[10] Narváez-García, A., Zavala-Loría, J.C., Ruiz-Marín, A. & Córdova-Quiroz, A.V.; *Simple Multicomponent Batch Distillation Procedure with a Variable Reflux Policy*; **2014**, 31 (2), pp. 531–542.

[11] Diwekar, U.M.; *Simulation, Design, Optimization, and Optimal Control of Multicomponent Batch Distillation Columns*; Ph. D. Thesis, Indian Institute of Technology, Bombay, India. **1988**.

[12] Diwekar, U. M. & Madhavan, K. P.; *Multicomponent Batch Distillation Column Design*; Industrial Engineering Chemistry. Research. **1991**, 30 (4), pp. 713–721.

[13] Sundaram, S. & Evans, B. L.; *Shortcut Procedure for Simulating Batch Distillation Operations*; Industrial Engineering Chemistry Research. **1993**, 32 (3), pp. 511–518.

[14] Shiras, R.N., Hanson, D.N. & Gibson, C.H.; *Calculation of Minimum Reflux in Distillation Columns*; Industrial Engineering Chemistry. **1950**, 42 (5), pp. 871–876.

[15] González-Velasco, J. R., Gutiérrez-Ortiz, M. A., Castrana-Pelayo, J. M. & González-Marcos, J. A.; *Improvements in Batch Distillation Startup*; Industrial Engineering Chemistry. **1987**, 26 (4), pp. 745–750.

[16] Zavala-Loría, J.C., Ruíz-Marín, A. & Coronado-Velasco, C.; *Maximum Thermodynamic Efficiency Problem in Batch Distillation*; Brazilian Journal of Chemical Engineering. **2011**, 28 (2), pp. 333–342.

[17] Domench, S. & Enjalbert, M.; *Program for Simulating Batch Rectification as a Unit Operation*, Computers and Chemical Engineering. **1981**, 5, p. 181.

[18] Fenske, M. R.; *Fractionation of Straight Run Pennssylvania Gasolina*; Industrial Engineering Chemistry. **1932**, 24 (5), pp. 482–485.

[19] Underwood, A. J. V.; *Fractional Distillation of Multicomponent Mixture*; Chemical Engineering Progress. **1948**, 44 (8), pp. 603–614.

[20] Gilliland, E. R.; *Multicomponent Rectification. Estimation of the Number of Theoretical Plates as a Function of Reflux*; Industrial Engineering Chemistry. **1940**, 32 (8), pp. 1102–1106.

[21] Eduljee, H. E.; *Equations Replace Gilliland Plot*; Hydrocarbon Processing. **1975**, 54 (9) pp. 120–122.

[22] Winn, W. F.; *New Relative Volatility Method for Distillation Calculations*; Petroleum Refiner. **1958**, 37 (5), pp. 2–6.

Chapter 3

Mathematical Modelling of Batch Distillation Columns: A Comparative Analysis of Non-Linear and Fuzzy Models

Adriana del Carmen Téllez-Anguiano,
Mario Heras-Cervantes, Juan Anzurez-Marín,
Gerardo Marx Chávez-Campos and
José Antonio Gutiérrez Gnecchi

Additional information is available at the end of the chapter

http://dx.doi.org/10.5772/66760s

Abstract

Distillation is the process most commonly used in industry to separate chemical mixtures; its applications range from cosmetic and pharmaceutical to petrochemical industries. The equipment required to perform the distillation process is known as distillation column. Since initial investment and maintenance costs for distillation columns are very high it is necessary to have an appropriate mathematical model that allows improving the comprehension of the column dynamics, especially its thermal behaviour, in order to enhance the control and safety of the process. This chapter presents a general panorama of the mathematical modelling of distillation columns, having as a specific case of study the comparison of a space-state non-linear model and a Takagi-Sugeno fuzzy model for a batch distillation column using a binary mixture (Ethanol-Water).

Keywords: mathematical modelling, distillation column, Takagi-Sugeno, non-linear models

1. Introduction

Distillation is the process most commonly used in industry to separate chemical mixtures, being the petrochemical industry one of the most important due to that oil distillation allows obtaining useful product, such as fuels. Distillation is also widely used in the pharmaceutical and cosmetics industry in order to obtain specific drugs and in the liquor industry to obtain wines and liquors, among other applications [1].

Distillation columns are the essential equipment required to perform the distillation process, these columns allow producing food, fuel, medicine, among other products. However, distillation columns represent an important investment in the process they are used, that is why it is necessary to have both, corrective and predictive maintenance, in order to prevent failures in the process as well as in the equipment.

Through the computational and technological continuous development, the industrial processes, such as distillation, have become very complex systems due to the high number of components they have and the several functions they develop, so their vulnerability has also increased. Having appropriate techniques to model distillation columns, such that these models allow implementing efficient and reliable control techniques, is very important to obtain the desired product quality, the adequate process functioning and to improve the security of the system and the user.

In the literature, different mathematical models have been used to improve distillation columns dynamics and comprehensions have been reported. Simple linear and non-linear models are representations that consider only few variables and low-order equations, simplifying the design and implementation of controllers using computational tools. Kienle [2] presents a low-order model for an ideal multicomponent distillation process considering the non-linear wave propagation theory.

Balasubramhanya and Doyle Iii [3] present a low-order model for a reactive multicomponent distillation column as well as the designing of a (MPC) predictive control to obtain the best quality of the distilled product. In Ref. [4], a model based on neural networks having the aim of optimizing the energy efficiency in a binary distillation column is presented. Lopez-Saucedo et al. [5] present the simulation and optimization of a model for a conventional and nonconventional batch distillation column.

Astorga et al. [6] and Cervantes et al. [7] present high-gain observers to estimate the light component composition in a continuous distillation column using a set of models for each plate of the column. In Ref. [8], a fault tolerant scheme for a distillation column, where observers are used to detect failures in the temperature sensors considering a non-linear model of the distillation column, is presented. The parametric identification is other methodology used to estimate certain variables in distillation columns as presented in Refs. [9, 10].

The Takagi-Sugeno fuzzy model is a useful tool to model and control complex systems based on the concept of decomposing a non-linear model in a multi-model structure formed by linear models not necessarily independents and fuzzy logic [11, 12], where the non-linear system representation is obtained through a weighted sum of all the sub-systems. The Takagi-Sugeno fuzzy model provides a solution to solve the designing and implementation issues in control strategies for non-linear systems, for instance, Wang et al. [13] propose a methodology to design control techniques for systems having a Takagi-Sugeno form.

The stability analysis of the Takagi-Sugeno fuzzy model can be solved considering the Lyapunov approach and by using the inner point tool as well as optimization techniques based on linear matrix inequalities (LMIs) [14].

http://dx.doi.org/10.5772/66760s

In this chapter, the design and simulation of a non-linear state-space and Takagi-Sugeno models for a batch distillation column are presented. These models are simulated and compared in order to analyse if they aim the objective of representing adequately the process dynamics in order to facilitate the implementation of control strategies to improve the distilled product quality as well as the process security.

2. Distillation column operation modes

Due to the variety of substances found in the nature and their different phases (mainly liquid and vapour), there exist different distillation operation modes in order to separate diverse mixtures, obtaining different quality of products.

The main distillation operation types are as follows:

- *Vacuum distillation:* A low-pressure system is used in order to obtain a low-temperature boiling of the substances in the mixture. Usually, a vacuum pump is used to generate the low-pressure state, as shown in **Figure 1a**.
- *Destructive distillation:* The substance is heated at high temperatures to be decomposed in other products that can be separated by fractionating, its operation is similar to the one used in wood and coal, as shown **Figure 1b**.
- *Extractive distillation:* Different separation agents are added to azeotropic mixtures, altering the relative volatility of the mixture components in order to allow their separation (see **Figure 1c**).
- *Fractionating distillation*: Liquid mixtures are separated by heating, considering a high heat exchange and the liquid and vapour molar rates. This distillation is used to separate composite mixtures/substances having different but close boiling temperatures. It usually considers a continuous operation, having a constant feeding flow through a feeding tray. The section above the feeding tray is named rectifying section, under the feeding tray is called stripping section, as shown in **Figure 2a**.
- *Batch distillation*: Widely used in industry when having small liquid quantities or when obtaining different products from a single mixture load is required. This operation does not have steady state due that the mixture composition varies in time; besides, it only allows enriching or rectifying the distilled (lighter) product (see **Figure 2b**).

In general, the different distillation operation modes have the same operating principle, mainly due the physical variables that interact in the process, such as temperature, composition, pressure and heating energy.

A typical distillation column is formed by a boiler, a condenser and *n* trays. The boiler is the element that provides the heating energy necessary to evaporate the mixture into it. The condenser provides the cooling necessary to condensate vapour, part of this vapour returns to the column to enrich the mixture, the rest is obtained as a distilled product. The column body is composed of a set of trays, where a partial separation of the mixture is performed due the circulation of liquid and vapour flow.

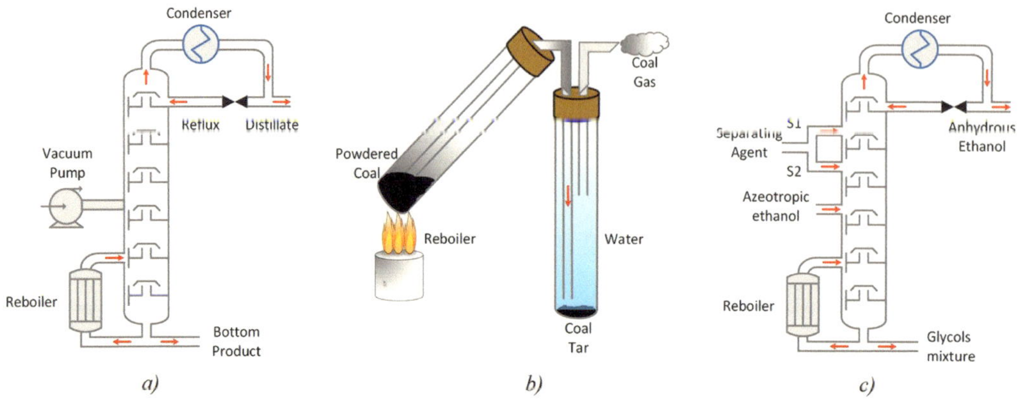

Figure 1. Distillation columns operation modes: (a) vacuum, (b) destructive and (c) extractive.

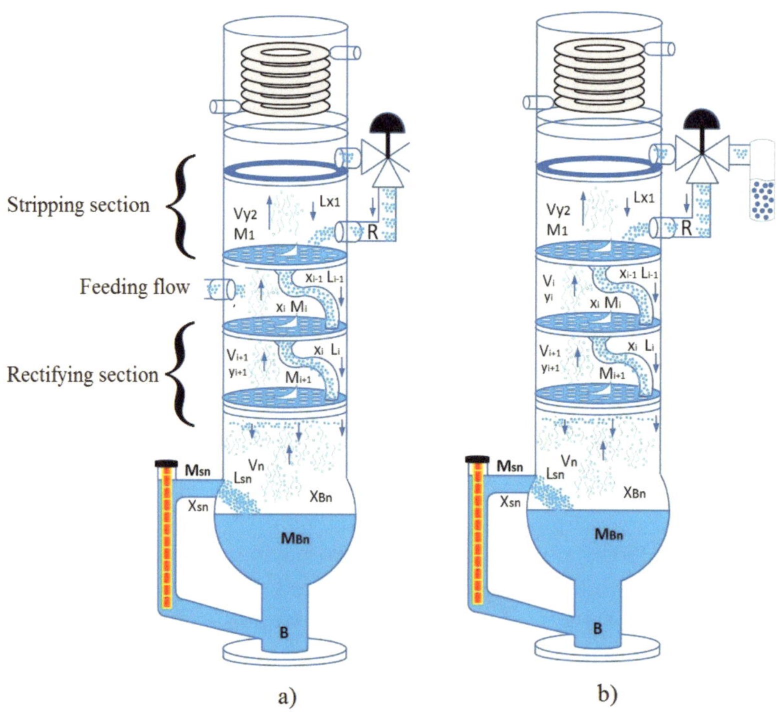

Figure 2. Distillation column operation modes: (a) fractionating and (b) batch.

The vapour flow is generated by the ebullition of the mixture in the boiler, the vapour rises into the column body and it is enriched by the light element of the mixture in each tray of the column. The liquid flow, generated by the reflux, descends from the condenser to the boiler by gravity and it is enriched by the heavy element of the mixture in every tray of the column. This operation can be described by an adequate mathematical model of the process.

http://dx.doi.org/10.5772/66760s

3. Distillation column mathematical modelling

The main objectives of designing a mathematical model of the distillation process are to simplify the analysis and comprehension of the distillation dynamics, facilitate the design of control techniques to enhance the distilled product quality and the system performance, estimate variables difficult to be measured, diagnose failures, among others. In order to deal with these objectives development of an adequate model is indispensable.

There exist different distillation column models according to its operation, the most commonly used in industrial applications are the continuous (fractionating) and batch models. Because of the similarity between the continuous and the batch operating types, in this section, a generic model that presents adequate results in both cases is presented.

It is well known that having a more complete/complex model implies having more complex equations difficult to solve, whereas having a simpler representation implies having simpler equations but the response resolution will have a higher calculation error compared to the real system response.

In general, there are two main model types according their complexity: simple and complex. The simple model found in the literature is the differential model, which considers the boiler and the condenser as trays in the distillation column. The column dynamics is represented by the component mass balance as shown in Eq. (1).

$$\frac{dWx_w}{dt} = \frac{Wdx_w}{dt} + \frac{x_w dW}{dt} = -Dy_D \tag{1}$$

where W is the bottom product, x_W is the bottom product composition, D is the distilled product and y_D is the distilled product composition.

The complex model considers each column element individually, i.e. a condenser, a boiler and trays are modelled individually, such that the response has a better resolution.

The particular study case presented in this chapter considers a complex model of a batch distillation column using a binary mixture.

4. Non-linear model of a binary batch distillation column

The model for a binary batch distillation column is obtained considering the light component, this component is obtained as a final (distilled) product [15]. The light component composition is obtained in each tray of the distillation column, where the liquid and vapour molar flows interact.

In order to design the distillation column model, the following assumptions are considered [16]: total condenser, no heating losses in the body column, constant pressure in the body column, liquid and vapour phases in thermodynamic equilibrium in each plate, variable relativity volatility according to the component composition.

The distillation column dynamics is represented by a set of differential equations that describe the behaviour of the light component of the mixture, given by Eq. (2).

$$\frac{dx_i}{dt} = \frac{L(x_{i\text{-}1} - x_i) + V(y_{i+1} - y_i)}{M_i} \tag{2}$$

where x_i is the liquid molar composition of the light component in tray i, y_i is the vapour molar composition of the light component in tray i, L is the liquid molar flow, V is the vapour molar flow and M is the retained mass.

The phase equilibrium is determined by constant K, as shown in Eq. (3) for ideal mixtures.

$$K = \frac{x_i}{y_i} \tag{3}$$

Such that considering the vapour-liquid equilibrium (VLE) and the relative volatility, the vapour composition as a function of the liquid composition is obtained. This function is presented in Eq. (4).

$$y = f(x, \alpha) \tag{4}$$

This is specifically presented in Eq. (5).

$$y_i = \frac{\alpha_i x_i}{1 + (\alpha_i - 1)x_i} \tag{5}$$

where α is the relative volatility in tray i.

Within each element of the distillation column flow different molar rates/quantities, named molar flows. These flows are the liquid and the vapour entering and leaving each tray, the distilled product and the bottom product.

In a binary batch distillation column, the liquid flows in both rectifying and stripping sections are the same, as well as the vapour flows, because there is not feeding flow.

$$\begin{aligned} L &= L_S = L_R \\ V &= V_S = V_R \end{aligned} \tag{6}$$

The molar flows considered in the binary batch distillation model are four: vapour (V), liquid (L), distilled (D) and bottom (B) products, these are expressed in Eqs. (7)–(9) [17].

$$V = \frac{Q_B}{H_1^{\text{vap}} x_n + H_1^{\text{vap}}(1 - x_n)} \tag{7}$$

where Q_B is the heating power (input), x_n is the liquid composition of light component in the boiler (tray n), H_1^{vap} is the vaporization enthalpy of the light component and H_2^{vap} is the vaporization enthalpy of the heavy component.

http://dx.doi.org/10.5772/66760s

$$L = (1\text{–}R)V \tag{8}$$

where R is the reflux input.

$$D = V\text{–}L \tag{9}$$

B, the bottom product, is not calculated, it is considered as the molar flow that remains into the boiler.

The non-linear model of the binary batch distillation column presented in this chapter is based on a set of sub-models, each sub-model corresponding to a specific element of the column (boiler, condenser and trays).

4.1. Condenser sub-model

The condenser is numbered as tray 1. Its dynamics is described by Eq. (10).

$$\frac{dx_1}{dt} = \frac{Vy_2\text{–}Lx_1\text{–}Dx_1}{M_1} \tag{10}$$

By substituting $L = (1\text{–}R)V$ in $D = V\text{–}L$, in order to represent the condenser as a function of the reflux, Eq. (11) is obtained.

$$D = RV \tag{11}$$

By substituting Eq. (11) in Eq. (10), Eq. (12) is obtained.

$$\frac{dx_1}{dt} = \frac{Vy_2\text{–}Lx_1\text{–}RVx_1}{M_1} \tag{12}$$

Considering that

$$y_2 = \frac{\alpha_2 x_2}{1\text{–}(\alpha_2\text{–}1)x_2} \tag{13}$$

the non-linear equation that represents the condenser dynamics is finally represented in Eq. (14).

$$\frac{dx_1}{dt} = \frac{V}{M_1}\left(\frac{\alpha_2 x_2}{1\text{–}(\alpha_2\text{–}1)x_2}\right) - \frac{Lx_1}{M_1} - \frac{RVx_1}{M_1} \tag{14}$$

4.2. Tray sub-model

The column body is formed by a set of n-2 trays. Eq. (15) describes its dynamics.

$$\frac{dx_i}{dt} = \frac{Vy_{i+2}\text{–}Vy_i + Lx_{i\text{-}1}\text{–}Lx_i}{M_i}; \quad i = 2, 3, \ldots, n\text{–}1 \tag{15}$$

where n is the total number of trays including a boiler and a condenser.

Considering that

$$y_i = \frac{\alpha_i x_i}{1-(\alpha_i-1)x_i} \tag{16}$$

the non-linear equation that represents the condenser dynamics is finally represented in Eq. (17).

$$\frac{dx_i}{dt} = \frac{V}{M_i}\left(\frac{\alpha_{i+1}x_{i+1}}{1-(\alpha_{i+1}-1)x_{i+1}}\right) - \frac{V}{M_i}\left(\frac{\alpha_i x_i}{1-(\alpha_i-1)x_i}\right) + \frac{L(x_{i-1}-x_i)}{M_i} \tag{17}$$

4.3. Boiler sub-model

The boiler is numbered as tray n. Eq. (18) describes its dynamics.

$$\frac{dx_n}{dt} = \frac{Vx_n - Vy_n + Lx_{n-1} - Lx_n}{M_n} \tag{18}$$

Factorizing Eq. (18), Eq. (19) is obtained.

$$\frac{dx_n}{dt} = \frac{V(x_n - y_n) + L(x_{n-1} - x_n)}{M_n} \tag{19}$$

Solving V to represent Eq. (19) as a function of the heating power (Q_B) based on Eq. (7), Eq. (20) is obtained.

$$\frac{dx_n}{dt} = \left(\frac{Q_B}{H_1^{\text{vap}}x_n + H_1^{\text{vap}}(1-x_n)}\right)\left(\frac{x_n - y_n}{M_n}\right) + \frac{L(x_{n-1}-x_n)}{M_n} \tag{20}$$

Then, considering

$$y_n = \frac{\alpha_n x_n}{1-(\alpha_n-1)x_n} \tag{21}$$

the non-linear equation that represents boiler dynamics is finally represented in Eq. (22).

$$\frac{dx_n}{dt} = \left[\left(\frac{Q_B}{H_1^{\text{vap}}x_n + H_1^{\text{vap}}(1-x_n)}\right)\left(\frac{x_n}{M_n}\right)\left(\frac{1-\alpha_n}{1-(\alpha_n-1)x_n}\right)\right] + \frac{L(x_{n-1}-x_n)}{M_n} \tag{22}$$

4.4. State-space non-linear model for a binary batch distillation column

In this section, the distillation column sub-models shown in Eqs. (14), (17) and (22) are presented in a state-space representation having the form shown in Eq. (23).

$$\dot{x} = Ax + Bu \tag{23}$$

This representation is used in a specific study case, a 12-tray distillation column including a boiler and a condenser, using a binary mixture in a batch operation. Compositions

$x = [x_1, x_2, \ldots, x_{12}]$ are considered as states of the model and $u = [R, Q_B]^T$ as inputs of the model. Matrices A and B are shown in Eqs. (24) and (25), respectively.

$$A = \begin{bmatrix} -\frac{L}{M_1} & \frac{Vf(x_2,\alpha_2)}{M_1} & 0 & 0 & \cdots & 0 & 0 \\ \frac{L}{M_2} & \frac{-L-Vf(x_2,\alpha_2)}{M_2} & \frac{Vf(x_3,\alpha_3)}{M_2} & 0 & \cdots & 0 & 0 \\ 0 & \frac{L}{M_3} & \frac{-L-Vf(x_3,\alpha_3)}{M_3} & \frac{Vf(x_4,\alpha_4)}{M_3} & \cdots & 0 & 0 \\ \vdots & \vdots & \vdots & \vdots & \ddots & \vdots & \vdots \\ 0 & 0 & 0 & 0 & \cdots & \frac{L}{M_{12}} & -\frac{L}{M_{12}} \end{bmatrix} \quad (24)$$

$$B = \begin{bmatrix} \frac{Vx_1}{M_1} & 0 \\ 0 & 0 \\ \vdots & \vdots \\ 0 & 0 \\ 0 & \frac{x_{12}-f(x_{12}-\alpha_{12})}{\left(H_1^{vap}x_{12} + H_1^{vap}(1-x_{12})\right)M_{12}} \end{bmatrix} \quad (25)$$

5. Takagi-Sugeno fuzzy model for a binary batch distillation column

The Takagi-Sugeno fuzzy representation describes the system dynamics based on linear sub-models interpolation and fuzzy rules [18].

Rule for model *j:*

If $z_1(t)$ is M_{1jj}, $z_2(t)$ is M_{2j},… and $z_p(t)$ is M_{pj}

Then:

$$x(t) = \sum_{j=1}^{r} A_j x(t) + B_j u(t) \quad (26)$$

where $j = 1, 2,\ldots, r$, M_j is the fuzzy set, r is the sub-model number, x is the state vector, u is the input vector, A_j is the state matrix for sub-model j, B_j is the input matrix for sub-model j and $z_j(t)$ is the scheduling measurable variable (state variables or external disturbances).

Given $[x(t,), u(t), z(t)]$, the complete fuzzy model is obtained by using a singleton-type fuzzifier, a product-type defuzzifier mechanism and the gravity centre. The Takagi-Sugeno fuzzy model for the non-linear system is expressed in Eq. (27).

$$\dot{x}(t) = \frac{\sum_{j=1}^{r} \omega_j \left(z_j(t) \right) [A_j x(t) + B_j u(t)]}{\sum_{j=1}^{r} \omega_j \left(z_j(t) \right)} \tag{27}$$

where the weight $\omega_j \left(z_j(t) \right)$ is 0 or a positive value, such that the sum of all the weights is positive; thus, the normalized weight, *hi*, is calculated in every rule from the z_j membership functions in the M_{jk} set. It is well known by fuzzy logic that $h_j = h_j[z(t)] \geq 0$ and $\sum_{j=1}^{r} h_j[z_j(t)] = 1$, as expressed in Eq. (28).

$$h_j[z_j(t)] = \frac{\omega_j \left(z_j(t) \right)}{\sum_{j=1}^{r} \omega_j \left(z_j(t) \right)} \tag{28}$$

The system expressed in Eq. (27) is equivalent to the system in Eq. (29).

$$\dot{x}(t) = \sum_{j=1}^{r} h_j [A_j x(t) + B_j u(t)] \tag{29}$$

5.1. Application to a binary batch distillation column

In this chapter, the specific study case is a 12-tray distillation column, including a boiler and a condenser, using an ethanol-water mixture in a batch operation. In the Takagi-Sugeno fuzzy model the liquid (L) and vapour (V) molar flows are proposed as parameters; the nominal operating ranges in steady state are:

$$\begin{aligned} L &= [0.418783, 2.97801] \\ V &= [0.418783, 2.97801] \end{aligned} \tag{30}$$

According to these parameters, the Takagi-Sugeno fuzzy model that interpolates between four linear models based on the following rules is obtained:

Rule 1:

$$\text{if } V \text{ is } V_{\min} \text{ and if } L \text{ is } L_{\min} \tag{31}$$

Then:

$$\dot{x}_1(t) = A_1 x(t) + B_1 u(t) \tag{32}$$

Rule 2:

$$\text{if } V \text{ is } V_{\min} \text{ and if } L \text{ is } L_{\max} \tag{33}$$

Then:

$$\dot{x}_2(t) = A_2 x(t) + B_1 u(t) \tag{34}$$

Rule 3:

$$\text{if } V \text{ is } V_{\max} \text{ and if } L \text{ is } L_{\min} \tag{35}$$

Then:

$$\dot{x}_3(t) = A_3 x(t) + B_2 u(t) \tag{36}$$

Rule 4:

$$\text{if } V \text{ is} V_{\max} \text{ and if } L \text{ is } L_{\max} \tag{37}$$

Then:

$$\dot{x}_4(t) = A_4 x(t) + B_2 u(t) \tag{38}$$

where:

$$\begin{aligned} A_1 &= \{V_{\min}, L_{\min}, G\{x_1, \alpha_1\}, \ldots, G\{x_{12}, \alpha_{12}\}, M_1, \ldots, M_{12}\} \\ A_2 &= \{V_{\min}, L_{\max}, G\{x_1, \alpha_1\}, \ldots, G\{x_{12}, \alpha_{12}\}, M_1, \ldots, M_{12}\} \\ A_3 &= \{V_{\max}, L_{\min}, G\{x_1, \alpha_1\}, \ldots, G\{x_{12}, \alpha_{12}\}, M_1, \ldots, M_{12}\} \\ A_4 &= \{V_{\max}, L_{\max}, G\{x_1, \alpha_1\}, \ldots, G\{x_{12}, \alpha_{12}\}, M_1, \ldots, M_{12}\} \end{aligned} \tag{39}$$

$$\begin{aligned} B_1 &= \begin{pmatrix} \dfrac{V_{\min} \cdot x_1}{M_1} & 0 \\ 0 & 0 \\ \vdots & \vdots \\ 0 & 0 \\ 0 & \dfrac{x_{12} - g(x_{12}, \alpha_{12})}{(H_{\mathrm{EtoH}}^{\mathrm{vap}} x_{12} + H_{\mathrm{H_2O}}^{\mathrm{vap}}(1 - x_{12})) \cdot M_{12}} \end{pmatrix} \\ B_2 &= \begin{pmatrix} \dfrac{V_{\max} \cdot x_1}{M_1} & 0 \\ 0 & 0 \\ \vdots & \vdots \\ 0 & 0 \\ 0 & \dfrac{x_{12} - g(x_{12}, \alpha_{12})}{(H_{\mathrm{EtoH}}^{\mathrm{vap}} x_{12} + H_{\mathrm{H_2O}}^{\mathrm{vap}}(1 - x_{12})) \cdot M_{12}} \end{pmatrix} \end{aligned} \tag{40}$$

The membership functions *($\mu(z)$)* for the fuzzy set are determined by:

Eq. (41) for vapour *V*:

$$\mu(V) \begin{cases} \mu V_{\min} = \dfrac{V_{\max} - V}{V_{\max} - V_{\min}} \\ \mu V_{\max} = 1 - \mu V_{\min} \end{cases} \tag{41}$$

Eq. (42) for liquid *L*:

$$\mu(V)\begin{cases} \mu L_{\min} = \dfrac{L_{\max}-L}{L_{\max}-L_{\min}} \\ \mu L_{\max} = 1-\mu L_{\min} \end{cases} \tag{42}$$

The normalized weights are given by Eq. (43):

$$\begin{cases} h_1(V,L) = \mu V_{\min}\mu L_{\min} \\ h_2(V,L) = \mu V_{\min}\mu L_{\max} \\ h_3(V,L) = \mu V_{\max}\mu L_{\min} \\ h_4(V,L) = \mu V_{\max}\mu L_{\max} \end{cases} \tag{43}$$

The Takagi-Sugeno fuzzy model proposed for the distillation column is represented in Eq. (44).

$$x(t) = \sum_{i=1}^{r} h_i(L,V)(A_i x(t) + B_i u(t)) \tag{44}$$

6. Models experimental validation and comparison

The Takagi-Sugeno fuzzy model is validated in Matlab by using experimental data from a 12-tray batch distillation column with variable reflux, using an ethanol-water mixture and considering the characteristics presented in **Table 1**.

Parameter	Value	Units
EtOH volume in boiler	2000	mL
H_20 volume in boiler	2000	mL
Process total pressure	637.42	mmHg

Table 1. Mixture initial parameters.

The initial molar composition of ethanol in the boiler is 0.2216, considering that the feed volume corresponds to 96%Vol ethanol.

The characteristics of the process inputs for the study case, the heating power (Q_B) and the reflux valve opening (R) are shown in **Table 2**.

Figure 3 presents the temperatures estimated by the Takagi-Sugeno model for the trays in the column body. The temperature increment and decrement due the reflux (R) action can be seen in all the trays.

Figure 4 presents the temperature graphics corresponding to the condenser (a) and to the boiler (b) in the non-linear and Takagi-Sugeno model. Temperature variations existing during the heating power (Q_B) and reflux changes (R) are shown. It can be seen that there exist a difference between the results obtained by both models due the reflux action, this difference is provoked by the fixed operating points for liquid and vapour flows in the Takagi-Sugeno model; however, this difference is small (less than 1.5%).

Input	Signal	Time
Q_B	Step 0–800 J	0 min
R	Total	0 min
Q_B	Step 800–1000 J	3.3 min
Q_B	Step 1000–1250 J	5.98 min
R	Pulse (ton = 6 s, toff = 6 s)	12.61 min
Q_B	Step 1250–1100 J	14.78 min
Q_B	Step 1100–950 J	17.15 min
Q_B	Step 950–1100 J	19.36 min
R	Total	23 min
Q_B	Step1100–1250 J	24.88 min

Table 2. Input parameters.

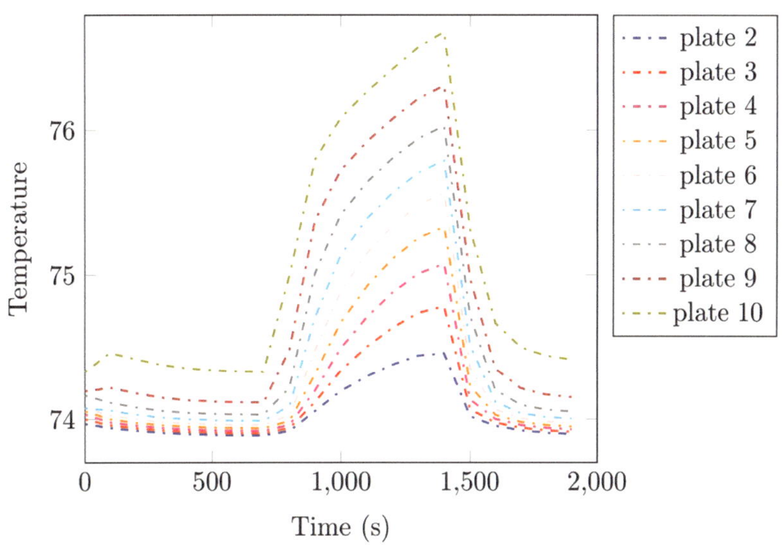

Figure 3. Plate temperatures in the distillation column.

In **Figure 5**, the composition graphics estimated for the distillation column trays by the Takagi-Sugeno fuzzy model are presented, these composition values vary according to the tray position.

In **Figure 6**, the simulation results obtained by the non-linear and Takagi-Sugeno models for the light component composition in the condenser (a) and the boiler (b) are presented. It can be seen that the composition behaviour in both trays varies according the heating power (Q_B) and reflux (R) changes, as shown in **Table 2**.

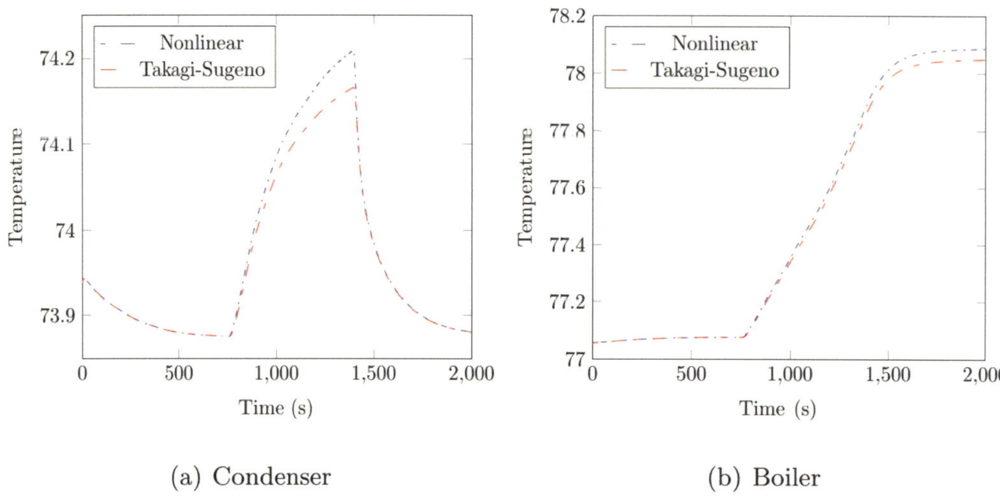

Figure 4. a) Condenser and b) Boiler temperatures, non-linear and Takagi-Sugeno models.

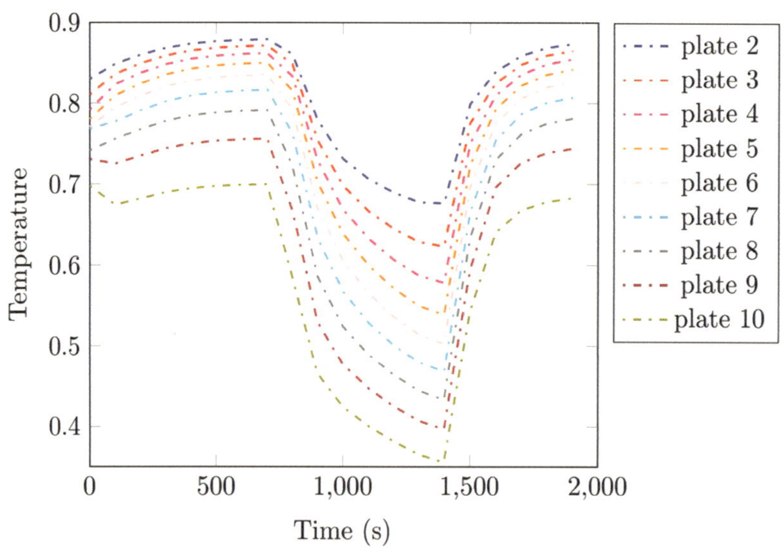

Figure 5. Plate temperatures in the distillation column.

Figure 7 shows the liquid and vapour molar flow behaviour during the distillation process. It can be seen the process dynamics when reflux or heating power changes exist.

The error percentage in the Takagi-Sugeno model compared to the non-linear models, calculated by the function shown in Eq. (45), is graphically represented in **Figure 8**. It can be seen that the error behaviour in the condenser (a) and the boiler (b) has a maximum value of 1.5% due to the reflux changes.

$$e = \frac{|x_{TS} - x_{NL}|}{x_{NL}} 100\% \tag{45}$$

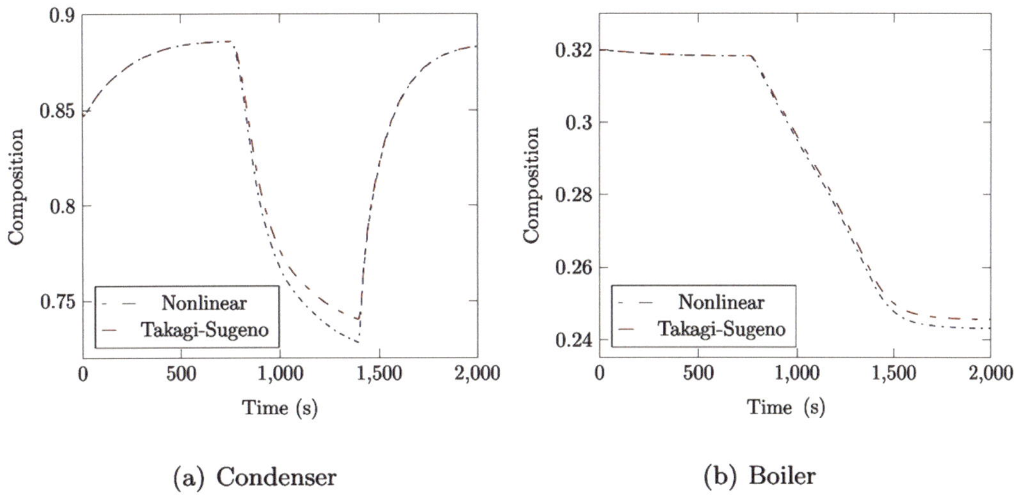

Figure 6. a) Condenser and b) Boiler temperatures obtained by non-linear and Takagi-Sugeno models.

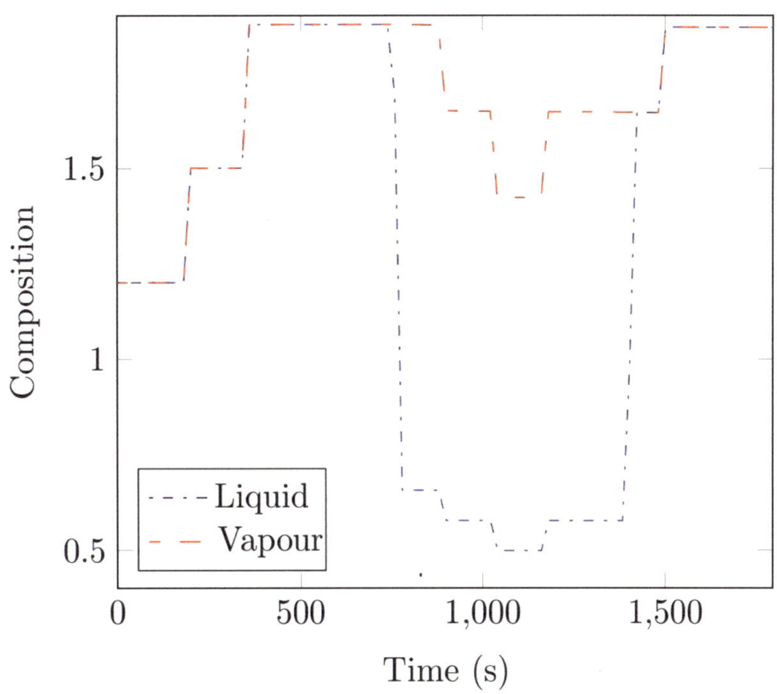

Figure 7. Liquid and vapour molar flows.

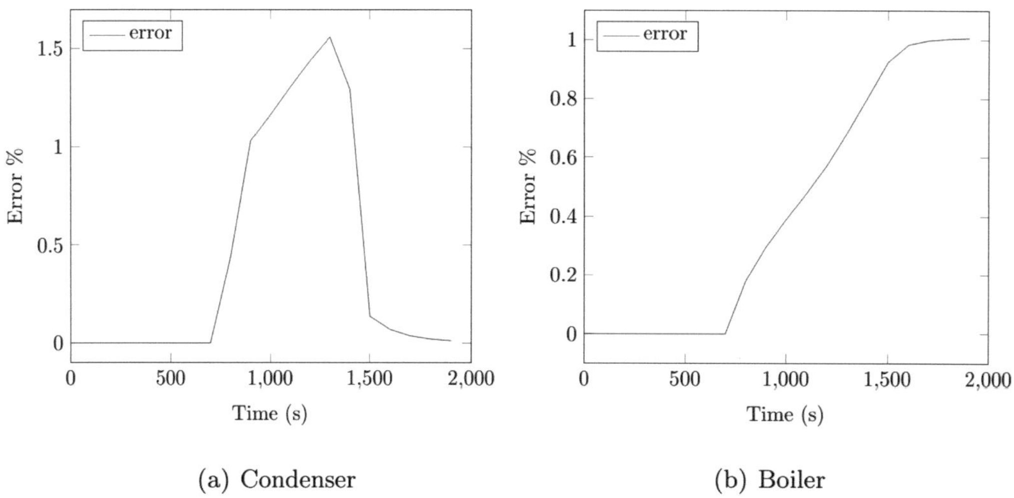

Figure 8. a) Condenser and b) Boiler error percentages.

7. Conclusions

This chapter presents the analysis and design of a state-space non-linear model and the Takagi-Sugeno fuzzy model for a batch distillation column using a binary mixture. The state-space non-linear model is based on differential equations considering compositions, temperatures and molar flows in the column. The linear fuzzy model is based on four rules, considering as parameters the liquid and vapour molar flows.

Both, the state-space non-linear and the linear fuzzy models are simulated in Matlab considering real input parameters (heating power and reflux) from a 12-tray batch distillation pilot plant using an ethanol-water mixture. The light component compositions and the temperatures in each tray of the column are calculated by both models. Besides, the obtained results are compared considering the same operating parameters, this comparison has the aim to verify the adequate functioning of the non-linear state-space and the Takagi-Sugeno models in order to analyse the existing differences.

The Takagi-Sugeno fuzzy model presents small differences in the estimations of the composition component and the tray temperatures when a reflux disturbance is presented due that the reflux affects directly the operating points established in this model; however, these differences are small enough to be neglected and both models converge under any operating condition.

The Takagi-Sugeno fuzzy model for a distillation column represents an alternative tool that takes advantage of the fuzzy control theory, allowing to facilitate the design and implement nonconventional control strategies for non-linear systems, however, if a higher resolution response is required it could be convenient to consider the non-linear model.

http://dx.doi.org/10.5772/66760s

Author details

Adriana del Carmen Téllez-Anguiano[1]*, Mario Heras-Cervantes[2], Juan Anzurez-Marín[2], Gerardo Marx Chávez-Campos[1] and José Antonio Gutiérrez Gnecchi[1]

*Address all correspondence to: actelleza@gmail.com

1 Electronics Department, Instituto Tecnológico de Morelia, Morelia, Michoacán, México

2 Universidad Michoacana de San Nicolas de Hidalgo, FIE, Morelia, Mich, México

References

[1] Mujtaba, I. M. Batch distillation [thesis]. London, England: Imperial College Press; 2004.

[2] Kienle, A. Low-order dynamic models for ideal multicomponent distillation processes using nonlinear wave propagation theory. Chemical Engineering Science. 2000;**55**(10):1817–1828.

[3] Balasubramhanya, L. S., & Doyle Iii, F. J. Nonlinear model-based control of a batch reactive distillation column. Journal of Process Control. 2000;**10**(2):209–218.

[4] Osuolale, F. N., & Zhang, J. Energy efficiency optimisation for distillation column using artificial neural network models. Energy. 2016;**106**:562–578.

[5] Lopez-Saucedo, E. S., Grossmann, I. E., Segovia-Hernandez, J. G., & Hernández, S. Rigorous modeling, simulation and optimization of a conventional and nonconventional batch reactive distillation column: A comparative study of dynamic optimization approaches. Chemical Engineering Research and Design. 2016;**111**:83–99.

[6] Astorga, C. M., Quintero-Mármol, E., Vela, L. G., & Torres, L. Observador No Lineal para la Estimación de Concentraciones en un Proceso de Destilación Metanol/Etanol. Información Tecnológica. 2006;**17**(6):35–41.

[7] Cervantes, M. H., Anguiano, A. T., Ramırez, M. G., Zaragoza, C. A., & Archundia, E. R. Validación experimental de un Observador de Alta Ganancia Reconfigurable para una columna de destilación. Revista Iberoamericana de Automática e Informática Industrial RIAI. 2015;**12**(4):397–407.

[8] Ortiz-Torres, G., Escobar, R. F., Adam-Medina, M., Astorga-Zaragoza, C. M., & Guerrero-Ramírez, G. V. Control Tolerante a Fallas Activo en Sensores y Actuadores: Aplicación a una Columna de Destilación. Revista Iberoamericana de Automática e Informática Industrial RIAI. 2016;**13**(1):67–79.

[9] Zhu, Y. Distillation column identification for control using Wiener model. In: American Control Conference, IEEE. San Diego, CA, USA 1999 . Proceedings of the 1999; 1999. pp. 3462–3466.

[10] Safdarnejad, S. M., Gallacher, J. R., & Hedengren, J. D. Dynamic parameter estimation and optimization for batch distillation. Computers & Chemical Engineering. 2016;**86**:18–32.

[11] Johanson, T. A., & Murray-Smith, R. Operating regime approach to nonlinear modeling and control. In: Multiple model approaches to modeling and control. Taylor and Francis, UK. 1997; pp. 3–72.

[12] Angelov, P. P., & Filev, D. P. An approach to online identification of Takagi-Sugeno fuzzy models. IEEE Transactions on Systems, Man and Cybernetics, Part B (Cybernetics). 2004;**34**(1):484–498.

[13] Wang, H. O., Tanaka, K., & Griffin, M. Parallel distributed compensation of nonlinear systems by Takagi-Sugeno fuzzy model. Fuzzy Systems, 1995. International Joint Conference of the Fourth IEEE International Conference on Fuzzy Systems and the Second International Fuzzy Engineering Symposium, Proceedings of 1995 IEEE, Yokohama, Japan. 1995;**2**:531–538.

[14] Tanaka, K., & Sugeno, M. Stability analysis and design of fuzzy control systems. Fuzzy Sets and Systems. 1992;**45**(2):135–156.

[15] Cingara, A., Jovanovic, M., & Mitrovic, M. Analytical first-order dynamic model of binary distillation column. Chemical Engineering Science. 1990;**45**(12):3585–3592.

[16] Skogestad, S. Dynamics and control of distillation columns-a critical survey. Modeling Identification and Control. Norwegian Society of Automatic Control. 1st ed. 1997.

[17] Aguilera Gonzalez, A., Tellez-Anguiano, A., Astorga-Zaragoza, C. M., Juárez-Romero, D., & Quintero-Mármol, E. Observador de alta ganancia constante para una clase de sistema no lineal de forma triangular. Revista Iberoamericana de Automática e Informática Industrial. 2010;**7**(2):31–38.

[18] Takagi, T., & Sugeno, M. Fuzzy identification of systems and its applications to modeling and control. IEEE Transactions on Systems, Man and Cybernetics. 1985;**1**:116–132.

Section 2

Distillation Applications

Chapter 4

Distillation: Basic Test in Quality Control of Automotive Fuels

Ma Mercedes del Coro Fernández-Feal,
Luis R. Sánchez-Fernández and
Blanca Sánchez-Fernández

Additional information is available at the end of the chapter

http://dx.doi.org/10.5772/67140

Abstract

The petroleum-derived automotive fuels available on the market today have different characteristics from those that were available a decade ago, mainly due to the promotion of the use of biofuels. However, the study of their distillation curves remains a basic test for their quality control. The ISO 3405 Standard has been the basis of the test procedure for the determination of the distillation characteristics of petroleum-derived automotive fuels at atmospheric pressure. This test is essential for the quality control of this type of products because of the extensive information that can be extracted from the interpretation of its results. The introduction of biofuels (bioethanol, biodiesel) in the new automotive fuel formulations, petrol and diesel fuels, made imperative to review ISO 3405:2000 in 2011. This paper studies the most significant changes between the two versions of the ISO 3405. The latest edition of the Standard is broader in scope; it has been modified in order to include the new fuel formulations which result from biofuel mixtures and the new criteria for repeatability and reproducibility calculation. This paper studies the most significant changes between the two versions of the ISO 3405 Standard together with a field study of commercial automotive fuel samples selection (with and without biofuel blend) and certified reference materials.

Keywords: distillation, volatility, quality control, fuel, biofuel

1. Introduction

'Distillation is the most widely used separation technique in the petroleum industry' [1].

At present, petroleum remains the major source of energy resources, and for more than 100 years, it has been the main source of fuels used in alternative internal combustion engines in auto-motion as well, both for spark ignition engines (SIE[1]), traditionally known as petrol engines, and for compression ignition engines (CIE[2]), or diesel engines. Nevertheless, we must bear in mind that differences in the operation of spark ignition and compression ignition engines require very different types of fuels.

When we speak about petroleum-derived fossil fuels, used as automotive fuel, we must remember that in a given series of hydrocarbons the ignition temperature decreases as the molecular weight increases because the cracking of large molecules needs less activation energies [1], whereby:

1. The SIE require low boiling hydrocarbons, with a soft combustion temperature and a relatively high spontaneous ignition temperature.
2. In the CIE, hydrocarbons with low spontaneous ignition temperatures are preferable, whereby the compounds of low boiling points are unsuitable.

From the chemical composition viewpoint, petrol is a blend of hydrocarbons between C4 and C11, with boiling points between 25 and 210°C and in which we can find all types of hydrocarbons: paraffins, isoparaffins, olefins, aromatics, naphthenes, etc. They may also contain oxygenated compounds such as ethers (MTBE, ETBE, TAME[3]) and pure alcohols in variable proportions: minimum amounts of sulphur and nitrogen as well as additives (detergents, anti-knock, etc.).

From the chemical composition viewpoint, the diesel fuels are a blend of different components obtained from different refining processes, with a majority of hydrocarbons between C10 and C16, with boiling ranges between about 160 and 360°C, and low amounts of sulphur and nitrogen. Additives are also present in their formulations [2–4].

Petroleum is, even today, the source of energy most used worldwide, but it is not an inexhaustible source of energy; this fact, together with the need to protect the environment, has led to the search for new automotive fuels and for the modification of the characteristics of the existing ones. One of the paths chosen in recent years has been to include biofuels in the formulation of conventional automotive fuels given their great capacity for blend with petroleum-derived fuels [5].

Incorporating biofuels and/or bioethanol to petrol and biodiesel (FAME[4]) to diesel is a path to a more sustainable energy future and involves a great R&D effort since it is necessary to study how blending modifies fuel characteristics in the search for optimal behaviour in any automobile engine [6].

As a general rule, according to international specifications, petrol can be combined with bioethanol in a percentage not exceeding 10% by volume and diesel with biodiesel up to a maximum of 7% in volume without informing consumers about it. There are also specially designed vehicles that can support a mixture of petrol with 85% by volume of bioethanol (E85) and taxis or buses that use a mixture of diesel and biodiesel in a proportion of 70:30% by volume (B30) [7].

Now, any fuel which is in the market should guarantee that their use in an engine will provide the projected energy performance that it will satisfy any other capacity inherent to their

[1]SIE—spark ignition engine
[2]CIE—compression ignition engine
[3]Methyl tert-butyl ether, ethyl tert-butyl ether, tert-amyl methyl ether
[4]FAME—fatty acid methyl esters

http://dx.doi.org/10.5772/67140

use and it will perform to the environmental quality level required. In order to comply with these conditions, those fuels must comply with certain specifications, i.e. a set of physical and chemical characteristics with maximum or minimum specified values, obtained through test procedures or standards [8, 9], including their volatility.

With the aim of controlling engine performance and the formation of vapours which may form explosive mixtures with air or escape to the atmosphere as emissions (VOCs—volatile organic compounds), most of the specifications for petroleum distillate products, specially the main automotive engine fuels, limit the values of certain distillation characteristics (volatility).

A fuel distillation range provides decisive information about its composition, its use and its behaviour during storage.

2. State of the art

2.1. Volatility versus fuel type

Among the wide variety of features to consider when establishing the quality of a fuel used in auto-motion, volatility stands out as one of the most critical ones since it is a characteristic directly related to engine performance and pollutant emissions [2–4].

2.1.1. Petrol

Petrol is a fuel which is a liquid state in the fuel tank and in the fuel injectors (or carburettor on older engines) and which is nebulized with air before being injected into the combustion chamber:

- If the volatility of fuel is low, the petrol does not exist in the gas phase, and there will be difficulties with the starting up of the engine and the behaviour of the engines in cold regimes.
- If the volatility is high, the petrol can be vaporized in the tank itself or in the pipelines ('vapour lock'). As a consequence the injection rate is inadequate, and the engine drowns.

2.1.2. Diesel

The volatility characteristics of a diesel have a great influence on the performance of diesel engines:

- If the volatility is low, then high distillation end points are obtained, which are indicative of high combustion times and poor combustion of heavy hydrocarbons. This will lead to the formation of smoke, loss of power and increased fuel consumption.
- If the volatility is high, then the fuel can cause incidents of 'vapour lock' in the lines.

2.2. Measure of volatility: distillation curve

Volatility is not a physical magnitude that can be measured directly; it is necessary to define methods of evaluating it. One universally used method to determine the volatility of a fuel is the distillation test [10, 11] that offers different information according to the type of fuel tested.

2.2.1. Petrol

The distillation test measures the percentage of vaporized fuel as the temperature increases. The test result is a curve obtained under standardized conditions of temperature versus percentage of evaporated fuel (**Figure 1**) wherein the different sections of the curve allow us to interpret the different behaviours of the product:

- At least 20% V/V fuel must be vaporized below 70°C to ensure good cold start capability. If this percentage is lower, then difficulties may occur at start-up; if it is too high, evaporation losses will occur in the fuel tank, and vapour bubbles may form in the intake manifold of the vehicle.
- The temperature at which 50% V/V of fuel is vaporized is a critical parameter, since if it is too low it can cause the solidification of water vapour contained in the intake air resulting in formation of ice on elements forming the blend.
- If the temperature at which 90% V/V of the fuel is vaporized is too high, then the fuel can remain in liquid form within the cylinder, displacing the lubricant and coming to cause oil dilution. Additionally, combustion may be hampered, causing irregular operation of the engine. Also, the presence of hydrocarbons of high boiling point in the petrol is decisive for the generation of polluting emissions.

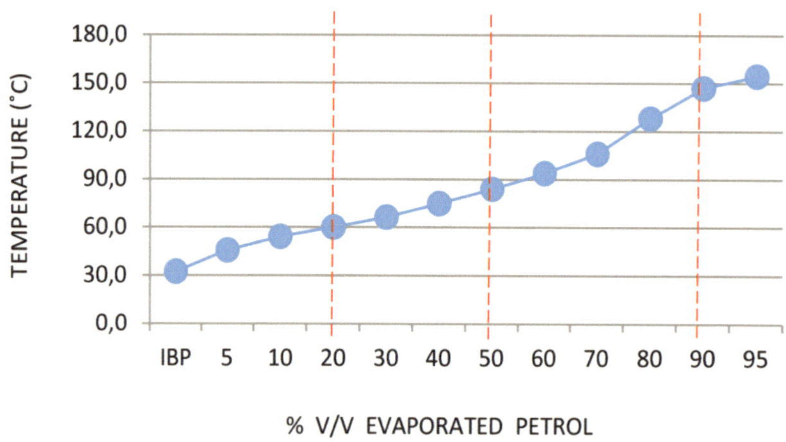

Figure 1. Distillation curve of petrol.

The percentages of fuel evaporated at 70, 100 and 150°C are limited. Additionally, a limitation in the final boiling point of to 210°C is also implemented in the regulations, ensuring the complete combustion of hydrocarbons and the non-formation of deposits in the combustion chamber and spark plugs.

2.2.2. Diesel

In a CIE, the volatility problems that may be present are notably different from those in SIE. In the regulations, there are no limitations in the light section of the curve but only for the end zone, where the fractionation of the components occurs.

The distillation test, in this case, measures the percentage of fuel which is recovered as the temperature increases. The test result is a curve obtained under standardized conditions of temperature versus percentage recovered (**Figure 2**). In these, fuel is important to note:

- The temperature at which vaporization ends, since if this is very high, combustion of the less volatile components will be incomplete, fuel droplets may reach the cylinder walls and dilution of the lubricating oil may take place, thereby increasing wear and producing coke deposits in the combustion chamber and waste segments [3].
- The end point of the fractionation of the components. This parameter is established so as 65% V/V must not have distilled before 250°C; 85% V/V should be distilled before 350°C,and 95% V/V before 360°C.

Figure 3 shows the distillation curves associated to both types of automotive fuels, petrol and diesel.

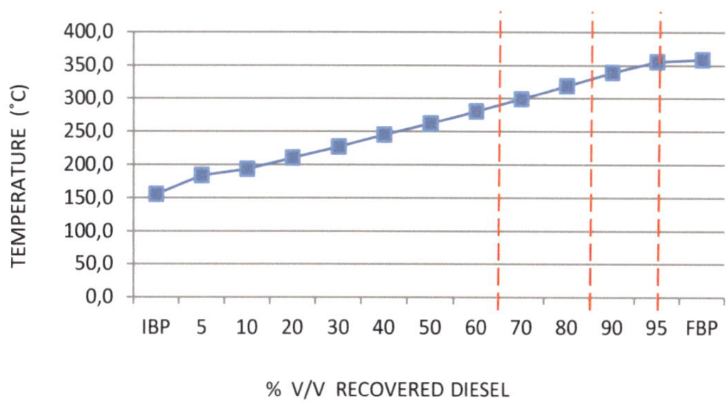

Figure 2. Distillation curve of diesel.

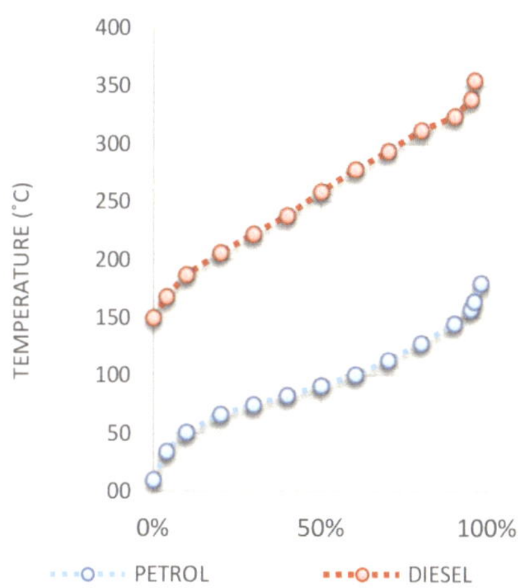

Figure 3. Comparison between distillation curves of petrol and diesel.

2.3. ISO 3405: petroleum products—determination of distillation characteristics at atmospheric pressure

ISO 3405 International Standard has been adopted by the European Standard EN ISO 3405, and it received the rank of the National Standard, by publication in the national language of an identical text under the responsibility of a member of CEN (European Committee for Standardization): UNE-EN ISO 3405 (Spain), DIN-EN ISO 3405 (Germany), etc.

This procedure aims to establish the steps '*for determining the distillation characteristics of light and middle distillates derived from petroleum having initial boiling points above 0°C and end points below approximately 400°C*'.

There are standards similar to ISO 3405 developed by other agencies with the same objective, such as ASTM D86: Standard Test Method for Distillation of Petroleum Products and Liquid Fuels at Atmospheric Pressure, whose first version dates from 1978, or IP 123: Petroleum Products—Determination of Distillation Characteristics at Atmospheric Pressure.

The test samples are classified into 'groups' based on their composition and the characteristics of expected volatility. In the case of fuels for automotive engines, petrol with up to 10% V/V ethanol is included in Group 1 and diesels with up to 20% V/V biodiesel in Group 4 (**Table 1**). Belonging to a particular group defines the setup of the equipment to be used and the condenser temperature. Finally, it determines the operation conditions to be used in the process of distillation by ISO 3405 (**Table 2**).

	Group 1	Group 4
Type of sample	Petrol	Diesel
Reid vapour pressure (kPa)	≥65.5	<65.5
Characteristics of expected volatility		
Initial distillation point, IDP (°C)	–	>100
Final boiling point, FBP (°C)	≤250	>250

Table 1. Sample type by group and expected volatility characteristics.

	Group 1	Group 4
Temperature of condensador bath (°C)	0–1	0–60
Temperature of medium around recovered receiving cylinder (°C)	13–18	±3 of charge
Time from the first heat to IBP (min)	5–10	5–15
Time from IBP to 5% (V/V) recovered (s)	60–100	–
Uniform average rate from 5% (V V) of recovered to 5 mL in the flask (mL/min)	4–5	4–5
Time from 5 mL residue in the flask to FBP (min)	≤5	≤5

Table 2. Test conditions.

During the testing process, a 100 mL test portion is distilled under the specified conditions appropriate to the fuel group, and systematic observations of thermometer readings and volumes of condensate recovered are made.

2.4. Precision ISO 3405: automated apparatus

The introduction of new formulations in automotive fuels made necessary the adaptation of the standard in effect at the time of their appearance (ISO 3405:2000), and, given the volatility characteristics of the new formulations, it was also necessary to validate the test method again based on new precision criteria.

The new version of the ISO 3405 standard (2011) establishes new criteria, based on the group to which it belongs the sample tested, to determine the validity of two results obtained under the same conditions by one operator using the same apparatus, in the same operating conditions, on the same day, on identical samples (repeatability, r) or obtained by different operators working in different laboratories on identical material test (reproducibility, R).

In the case of petrol and bioethanol blends (Group 1), these criteria are listed in **Tables 3** and **4** and in the case of diesel and biodiesel blends (Group 4) in **Tables 5** and **6**.

2000	**Evaporated % (V/V)**	**2011**	
Repeatability Group 1		**Repeatability Group 1**	**Valid range (°C)**
3.9	IBP	0.0295 (E + 51.19)	20–70
$r_2 + 0.56$	10	1.33	35–95
r_2	50	0.74	65–220
r_2	90	0.00755 (E + 59.77)	110–245
4.4	FBP	3.33	135–260

r_2 is a constant function of the slope, ΔC/ΔV, at each distillation point, with values calculated from $r_2 = 0.673\,(\Delta C/\Delta V) + 1.131$.
E is the temperature at the percentage evaporated within the prescribed valid range.

Table 3. Repeatability (Group 1).

2000	**Evaporated % (V/V)**	**2011**	
Repeatability Group 4		**Repeatability Group 4**	**Valid range (°C)**
3.5	IBP	0.018 T	145–220
1.42 (ΔC/ΔV) + 1.2	10	0.0094 T	160–265
1.42 (ΔC/ΔV) + 1.2	50	0.94	170–295
1.08 (ΔC/ΔV) + 1.1	90	0.0041 T	180–340
1.08 (ΔC/ΔV) + 1.1	95	0.01515 (T–140)	260–340
3.5	FBP	2.2	195–365

Slope: ΔC/ΔV
T is the temperature at the percentage recovered within the prescribed valid range.

Table 4. Repeatability (Group 4).

2000	Evaporated % (V/V)	2011	
Reproducibility Group 1		Reproducibility Group 1	Valid range (°C)
7.2	IBP	0.0595 (E + 51.19)	20–70
$R_2 + 0.72$	10	3.20	35–95
R_2	50	1.88	65–220
$R_2 - 1.90$	90	0.019 (E + 59.77)	110–245
8.9	FBP	6.78	135–260

R_2 is a constant function of the slope, ΔC/ΔV, at each distillation point, with values calculated from $R_2 = 1.998\ (\Delta C/\Delta V) + 2.617$.

E is the temperature at the percentage evaporated within the prescribed valid range.

Table 5. Reproducibility (Group 1).

2000	Evaporated % (V/V)	2011	
Reproducibility Group 4		Reproducibility Group 4	Valid range (°C)
8.5	IBP	0.055 T	145–220
2.64 (ΔC/ΔV) + 3.0	10	0.022 T	160–265
3.97 (ΔC/ΔV) + 2.9	50	2.97	170–295
2.53 (ΔC/ΔV) + 2.0	90	0.015 T	180–340
2.53 (ΔC/ΔV) + 2.0	95	0.04227 (T–140)	260–340
10.5	FBP	7.1	195–365

Slope: ΔC/ΔV

T is the temperature at the percentage recovered within the prescribed valid range.

Table 6. Reproducibility (Group 4).

3. Materials and methods

Step 1.

A sample selection of automotive fuel (petrol, diesel) with and without blend of biofuel (bioethanol, biodiesel) for a field study is used. The variations in product volatility as a result of the presence of biofuel in its composition are checked.

Each sample is tested in duplicate by two operators.

Step 2.

The changes in criteria established for repeatability and reproducibility in Standard ISO 3405 are studied; certified reference materials are used.

3.1. Test description

3.1.1. Reagents and materials

- Acetone: cleaning solvent.
- Certified reference material (CRM).
- Distillation flasks: flasks should have a capacity of 125 mL and be constructed of heat-resistant glass, according to the dimensions and tolerances given in Standard ISO 3405:2011.
- Receiving cylinder: graduate cylinder of 100 mL capacity, with a mark at 100 mL and metal base.
- Residue cylinder of 5 mL capacity.
- Certified temperature-sensor pt100 to an accuracy of 0.01C.
- Centring device, for centring the temperature sensor, adjusts the neck distillation flask, allows to centre the temperature sensor and prevents steam leaks.

3.1.2. Apparatus

- Automated equipment for petroleum product distillation satisfies the requirements established in the Standard ISO 3405:2011; the maximum error tracking device level is 0.3 mL.

3.1.3. Preparation of apparatus

- Clean the condenser tube, employing acetone. Dry thoroughly to remove any portion of acetone used for cleaning the device.
- Check that the temperature probe is properly seated in the centring device. Check your state and proper cleaning.
- Choose the support plate of the flask as the orifice diameter thereof according to the type of sample to be tested (**Table 7**).
- Check that the value of atmospheric pressure recorded by the apparatus is coincident with that indicated by the recording barometer atmospheric pressure in the laboratory.

	Group 1	Group 4
Diameter of hole in flask-support board (mm)	38	50
Temperature at start of test (°C)		
Flask and thermometer	13–18	≤ambient
Flask-support board and shield	≤ambient	–
Receiving cylinder and sample	13–18	13 at ambient

Table 7. Preparation of apparatus.

3.1.4. Procedure

- Measure the test portion precisely to the 100 mL mark of the receiving cylinder, and then transfer it as completely as practical to the distillation flask, taking care that none of the liquid flows into the vapour tube. If irregular boiling (bumping) is expected, add a small volume of clean and dry boiling chips to the test portion.
- Fit the flask vapour tube, provided with a silicone rubber stopper, tightly into the condenser tube. Adjust the distillation flask in a vertical position so that the vapour tube extends into the condenser tube for a distance of 25–50 mm. Raise and adjust the flask-support board to fit snugly against the bottom of the flask.
- Fit the receiving cylinder with a drip deflector, through which the distillate is going to drip.
- Place the receiving cylinder that was used to measure the test portion, without drying, into the bad under the lower end of the condenser tube so that the end of the condenser tube is centred in the receiving cylinder and extends therein for a distance of at least 25 mm.

After the choice of the method, which has been specifically developed to the distillation test according to **Table 2** and the corresponding group, follow the steps indicated by the system software. Any distillation that does not meet above conditions must be repeated, as well as those in which the actual loss differs by more than 2 mL from the estimated value.

3.1.5. Calculations

The data required for calculations is recorded in the range between the initial and the final boiling point, with an accuracy of 0.1 mL for all the readings on the receiving cylinder and with an accuracy of 0.1°C for all the readings on the temperature sensor.

Distillations carried out with automated instrument do not require manual calculation; the system software makes the appropriate calculations according to the corrective measures established by the Standard. However, it is certainly right to check the atmospheric pressure value does not differ by more than ±10 hPa from the value provided by the barometer in the laboratory.

3.2. Results obtained

3.2.1. Step 1: Samples of petrol—distillation curves obtained

Distillation test is carried out on 24 petrol samples with a bioethanol percentage that varies between 0.0 and 4.0% V/V, obtaining their corresponding distillation curves. Among these curves, we have selected seven for being analyzed and included in the present paper. Obtained results are shown in **Figure 4**.

After studying the trend and as the different sections of the curve give the opportunity to interpret the different product performances, a particular observation of each section is developed, making a graphic comparison on the basis of an average value (**Figures 5–7**).

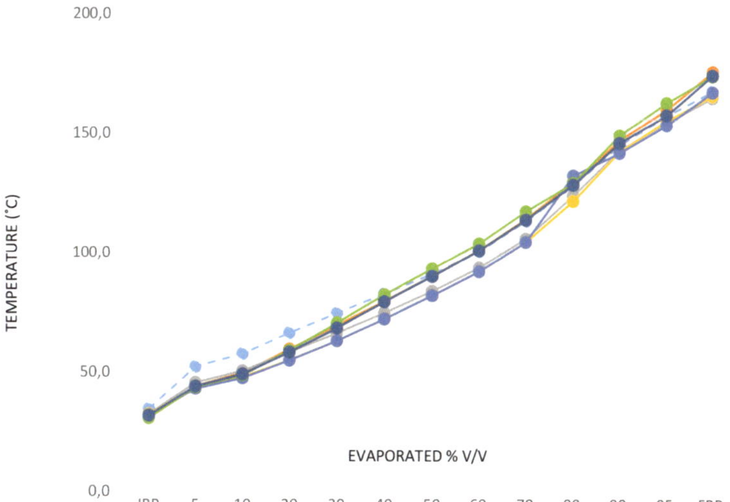

	IBP	5	10	20	30	40	50	60	70	80	90	95	FBP
0,0%V/V	34,4	52,2	57,6	66,3	74,6	82,7	91,0	100,5	113,2	128,1	145,0	157,4	167,7
1,0%V/V	31,2	44,1	49,6	59,5	69,4	79,6	90,1	101,0	114,1	129,6	147,0	160,0	175,8
1,5%V/V	32,3	45,6	50,5	58,3	66,2	74,7	83,8	93,7	105,9	123,9	142,8	154,5	164,9
2,1%V/V	32,0	43,9	47,8	55,0	63,0	72,1	82,0	92,0	104,4	121,6	142,2	155,3	165,7
2,5%V/V	31,5	42,9	47,4	54,8	63,0	72,1	81,9	92,0	104,4	132,3	141,7	153,4	167,3
3,5%V/V	30,6	43,6	48,7	58,9	70,5	82,3	93,2	103,8	117,3	129,1	149,1	162,9	174,0
4,0 %V/V	31,7	44,0	49,1	58,2	68,3	79,3	90,0	100,8	113,7	128,4	145,9	157,6	174,3

Figure 4. Distillation curves: petrol and blends with bioethanol (0.0–4.0% V/V).

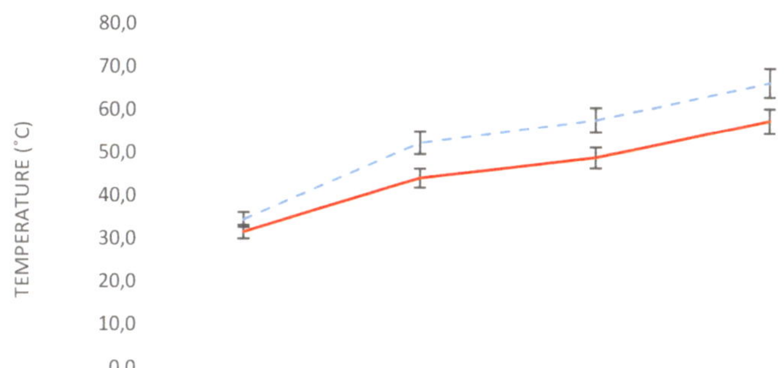

	IBP	E5	E10	E20
PETROL	34,4	52,2	57,6	66,3
PETROL WITH BIOETHANOL	31,6	44,0	48,9	57,5

Figure 5. Initial section (IBP-E20). Distillation curves of petrol samples.

3.2.2. Step 1: Diesel samples—distillation curves obtained

Distillation test is carried out on 24 diesel samples which contain a different biodiesel percentage (FAME), obtaining their corresponding distillation curves. Among these curves, we have selected seven for being analyzed and included in the present paper; six of them correspond to diesel samples with a usual biodiesel percentage, between 0.0 and 7.0% V/V. However, the other one corresponds to a diesel sample with a 30% V/V biodiesel percentage; this is outside the Standard scope. Obtained results are shown in **Figure 8**.

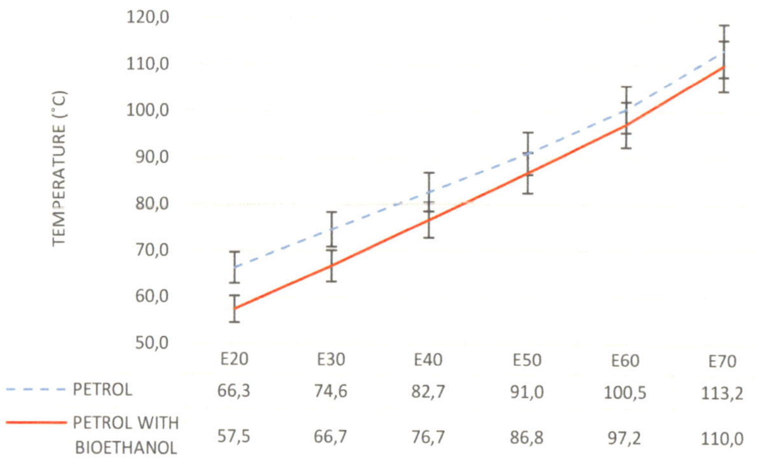

Figure 6. Middle section (E20-E70). Distillation curves of petrol samples.

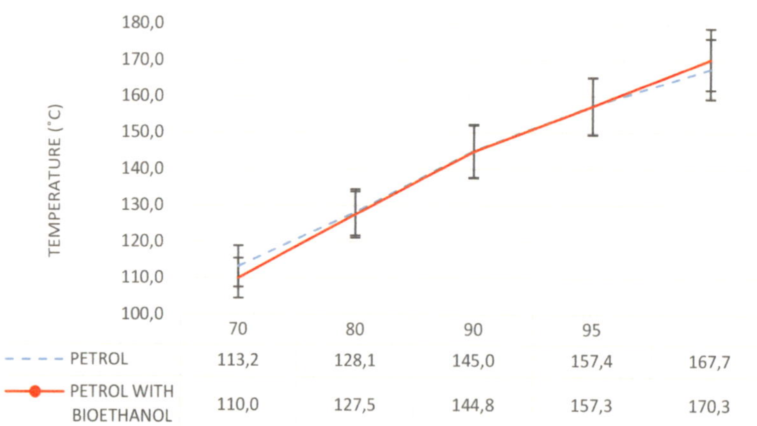

Figure 7. Final section (E70-FBP). Distillation curves of petrol samples.

As in the case of petrol samples, here we study the trend and the different sections of the distillation curve. However, in this case, the distillation curve will be divided for its study in two different sections: the first one comprises from the initial distillation point until 60% of the total volume is collected, whereas the second one comprises from this point to the final distillation point. A graphic comparison on the basis of an average value is also used here to illustrate the results, making a graphic comparison on the basis of an average value (**Figures 9** and **10**).

http://dx.doi.org/10.5772/67140

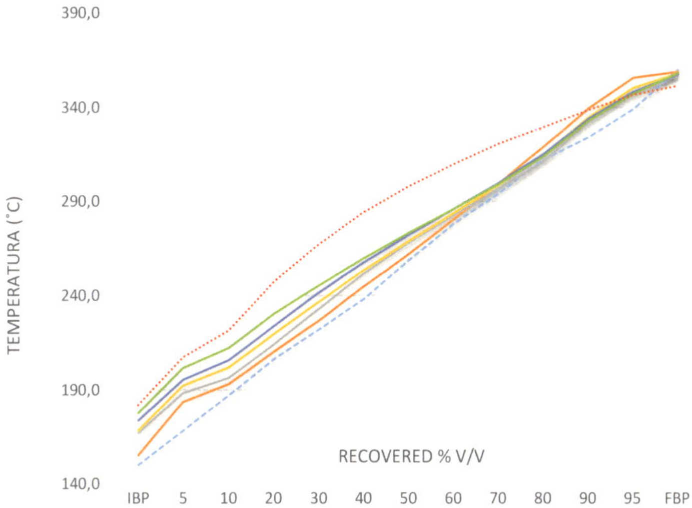

	IBP	5	10	20	30	40	50	60	70	80	90	95	FBP
0,0%	150,0	168,2	187,0	206,0	222,0	238,0	258,5	278,0	294,0	312,0	324,0	339,0	360,0
1,0 %	155,2	183,5	193,0	210,0	226,7	244,7	261,9	280,4	299,2	318,8	339,4	355,6	358,8
3,0%	167,0	188,2	196,3	213,9	232,9	251,5	268,1	282,7	296,1	311,0	332,1	350,1	356,9
4,0 %	168,4	192,1	201,8	219,4	236,7	253,4	268,9	283,7	298,5	314,6	334,6	350,2	357,8
5,0 %	173,6	195,3	205,7	223,6	241,6	257,6	272,1	285,9	299,6	314,9	333,9	348,1	357,5
7,0 %	177,6	201,5	212,2	230,1	245,4	259,8	273,3	285,9	299,0	313,6	332,8	347,1	358,1
30,0%	181,9	207,2	221,4	247,1	267,3	284,2	297,9	309,6	320,4	329,4	338,6	346,6	351,5

Figure 8. Distillation curves: diesel and blends with FAME (0.0–30.0% V/V).

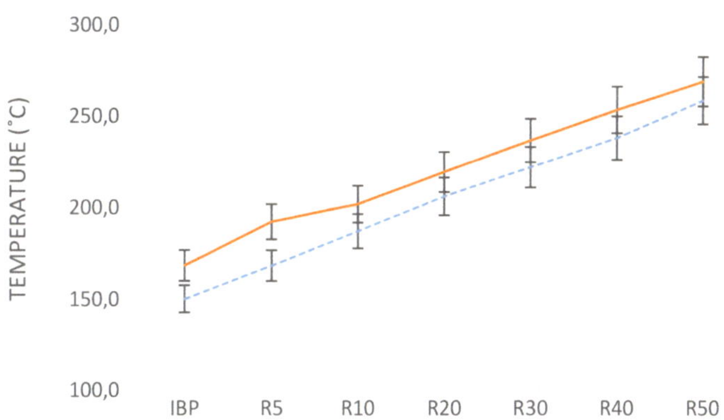

	IBP	R5	R10	R20	R30	R40	R50
DIESEL	150,0	168,2	187,0	206,0	222,0	238,0	258,5
DIESEL WITH FAME	168,4	192,1	201,8	219,4	236,7	253,4	268,9

Figure 9. Initial section (IBP-R50). Distillation curves of diesel samples.

3.2.3. *Step 2: Repeatability and reproducibility*

In order to know and be able to check the quantitative change derived from the precision requirement set in the new edition of the Standard, which involves repeatability and reproducibility, two different certified reference materials are used; one of them matches with the petrol boiling range, whereas the other one matches with the diesel boiling range.

Two analysts are involved in the test process; each using a different distillation automated equipment, in the same laboratory, and maintaining the required operation conditions for the standardized calculation. Distillation test is carried out in duplicate, and their results are shown in **Figures 11–14**.

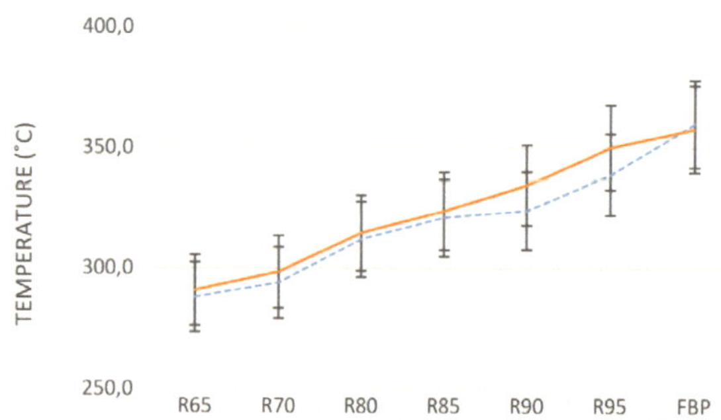

	R65	R70	R80	R85	R90	R95	FBP
DIESEL	288,0	294,0	312,0	321,0	324,0	339,0	360,0
DIESEL WITH FAME	290,9	298,5	314,6	323,8	334,6	350,2	357,8

Figure 10. Final section (R65-FBP). Distillation curves of diesel samples.

QUALITY CONTROL - PETROL - CRM - ISO 3405:2000

EVAPORATED %V/V	CERTIFICATE VALUE	ANALYST	TEMPERATURE (°C) MEASURED VALUES	AVERAGE OF ANALYST	DIFFERENCE BETWEEN AVERAGE OF ANALYST AND CERTIFICATE VALUE	AVERAGE OF ANALYSTS	DIFFERENCE BETWEEN AVERAGE OF ANALYSTS AND CERTIFICATE VALUE	TOLERANCE RANGE SIMPLE	DIFFERENCE BETWEEN VALUES BY ANALYST	REPEATABILITY	DIFFERENCE BETWEEN VALUES BY ANALYSTS	REPRODUCIBILITY
PI	32,2	1	33,2	33,80	1,20	32,68	1,60	5,09	1,20	3,90	3,10	7,2
		1	34,4									
		2	31,8	31,55	0,50		0,65		0,50			
		2	31,3									
10%	49,3	1	49,0	49,20	0,40	49,00	0,10	4,67	0,40	2,79	0,60	6,60
		1	49,4									
		2	48,8	48,80	0,00		0,50		0,00			
		2	48,8									
50%	83,7	1	83,6	83,70	0,20	83,73	0,00	3,08	0,20	1,72	0,20	4,35
		1	83,8									
		2	83,7	83,75	0,10		0,05		0,10			
		2	83,8									
90%	143,9	1	144,2	144,00	0,40	144,00	0,10	2,64	0,40	2,15	0,40	3,73
		1	143,8									
		2	143,8	144,00	0,40		0,10		0,40			
		2	144,2									
PF	166,3	1	165,8	165,95	0,30	165,93	0,35	6,29	0,30	4,40	0,30	8,90
		1	166,1									
		2	166,0	165,90	0,20		0,40		0,20			
		2	165,8									

Figure 11. Quality control according to ISO 3405:2000; petrol-CRM.

EVAPORATED %V/V	CERTIFICATE VALUE	ANALYST	TEMPERATURE (°C) MEASURED VALUES	AVERAGE OF ANALYST	DIFFERENCE BETWEEN AVERAGE OF ANALYST AND CERTIFICATE VALUE	AVERAGE OF ANALYSTS	DIFFERENCE BETWEEN AVERAGE OF ANALYSTS AND CERTIFICATE VALUE	TOLERANCE RANGE SIMPLE	DIFFERENCE BETWEEN VALUES BY ANALYST	REPEATABILITY	DIFFERENCE BETWEEN VALUES BY ANALYSTS	REPRODUCIBILITY
PI	32,2	1	33,2	33,80	1,20	32,68	1,60	3,51	1,20	2,46	3,10	5,0
		1	34,4									
		2	31,8	31,55	0,50		0,65		0,50			
		2	31,3									
10%	49,3	1	49,0	49,20	0,40	49,00	0,10	2,26	0,40	1,33	0,60	3,20
		1	49,4									
		2	48,8	48,80	0,00		0,50		0,00			
		2	48,8									
50%	83,7	1	83,6	83,70	0,20	83,73	0,00	1,33	0,20	0,74	0,20	1,88
		1	83,8									
		2	83,7	83,75	0,10		0,05		0,10			
		2	83,8									
90%	143,9	1	144,2	144,00	0,40	144,00	0,10	2,74	0,40	1,54	0,40	3,87
		1	143,8									
		2	143,8	144,00	0,40		0,10		0,40			
		2	144,2									
PF	166,3	1	165,8	165,95	0,30	165,93	0,35	4,79	0,30	3,33	0,30	6,78
		1	166,1									
		2	166,0	165,90	0,20		0,40		0,20			
		2	165,8									

Figure 12. Quality control ISO 3405:2011; petrol-CRM.

QUALITY CONTROL - DIESEL - CRM - ISO 3405:2000

RECOVERED %V/V	CERTIFICATE VALUE	ANALYST	TEMPERATURE (°C) MEASURED VALUES	AVERAGE BY ANALYST	DIFFERENCE BETWEEN AVERAGE OF ANALYST AND CERTEFICATE VALUE	AVERAGE BY ANALISTS	DIFFERENCE BETWEEN AVERAGE OF ANALYSTS AND CERTEFICATE VALUE	TOLERANCE RANGE SIMPLE	DFFERENCE BETWEEN VALUES BY ANALYST	REPEATABILITY	DFFERENCE BETWEEN VALUES BY ANALYSTS	REPRODUCIBILITY
PI	172,9	1	178,1	179,35	6,45	178,63	4,40	6,01	2,50	3,50	4,40	8,50
		1	180,6									
		2	176,2	177,90	5,00				3,40			
		2	179,6									
10%	209,2	1	210,4	210,25	1,05	209,98	1,20	5,49	0,30	3,76	1,20	7,76
		1	210,1									
		2	209,2	209,70	0,50				1,00			
		2	210,2									
50%	274,4	1	274,9	275,00	0,60	274,48	1,40	6,19	0,20	3,30	1,40	8,76
		1	275,1									
		2	273,7	273,95	0,45				0,50			
		2	274,2									
90%	339,2	1	341,3	341,05	1,85	339,78	3,40	4,93	0,50	3,22	3,40	6,97
		1	340,8									
		2	337,9	338,50	0,70				1,20			
		2	339,1									
95%	354,9	1	356,2	355,30	0,40	353,55	4,80	6,34	1,80	4,08	4,80	8,97
		1	354,4									
		2	351,4	351,80	3,10				0,80			
		2	352,2									
PF	364,2	1	359,4	359,60	4,60	359,60	3,20	7,42	0,40	3,50	3,20	10,50
		1	359,8									
		2	358,0	359,60	4,60				3,20			
		2	361,2									

Figure 13. Quality control according to ISO 3405:2000; diesel-CRM.

QUALITY CONTROL - DIESEL - CRM - ISO 3405:2011

RECOVERED %V/V	CERTIFICATE VALUE	ANALYST	TEMPERATURE (°C) MEASURED VALUES	AVERAGE OF ANALYST	DIFFERENCE BETWEEN AVERAGE OF ANALYST AND CERTIFICATE VALUE	AVERAGE OF ANALYSTS	DIFFERENCE BETWEEN AVERAGE OF ANALYSTS AND CERTIFICATE VALUE	TOLERANCE RANGE SIMPLE	DIFFERENCE BETWEEN VALUES BY ANALYST	REPEATABILITY	DIFFERENCE BETWEEN VALUES BY ANALYSTS	REPRODUCIBILITY
PI	172,9	1	178,1	179,4	6,45	178,63	5,72	6,72	2,50	3,11	4,40	9,51
		1	180,6									
		2	176,2	177,9	5,00				3,40			
		2	179,6									
10%	209,2	1	210,4	210,3	1,05	209,98	0,78	3,25	0,30	1,97	1,20	4,60
		1	210,1									
		2	209,2	209,7	0,50				1,00			
		2	210,2									
50%	274,2	1	274,9	275,0	0,80	274,48	0,28	2,10	0,20	0,94	1,40	2,97
		1	275,1									
		2	273,7	274,0	0,25				0,50			
		2	274,2									
90%	339,2	1	341,3	341,1	1,85	339,78	0,57	3,60	0,50	1,39	3,40	5,09
		1	340,8									
		2	337,9	338,5	0,70				1,20			
		2	339,1									
95%	354,9	1	356,2	355,3	0,40	353,55	1,35	6,42	1,80	3,26	4,80	9,08
		1	354,4									
		2	351,4	351,8	3,10				0,80			
		2	352,2									
PF	364,2	1	359,4	359,6	4,60	359,60	4,60	5,02	0,40	2,20	3,20	7,10
		1	359,8									
		2	358,0	359,6	4,60				3,20			
		2	361,2									

Figure 14. Quality control according to ISO 3405:2011; diesel-CRM.

4. Discussion

4.1. ISO 3405:2011 versus ISO 3405:2000

The fourth edition of ISO 3405 Standard came into force on 15 January 2011. This fourth edition cancelled and replaced the third edition, which had come into force on 1 March 2000. It is noteworthy that the text of this new edition is in line with the American Standard ASTM D86, widely used throughout the petroleum industry.

The fourth edition of ISO 3405 Standard introduces significant changes which are the result of the new formulations of automotive fuels and which are listed below:

The latest edition of the Standard is wider in scope. It has been modified in order to include the new fuel formulations, obtained from mixtures of biofuels and petrol (containing up to 10% V/V of ethanol) and diesel (containing up to 20% V/V of biodiesel). Additionally, it establishes a clear definition of 'light distillates' and 'middle distillates'.

Petrol from direct distillation (the petroleum fraction obtained by distillation at atmospheric pressure) no longer has their own group (the former Group 0, since the latest edition of the

http://dx.doi.org/10.5772/67140

Standard refers only to four distillation Groups 1, 2, 3 and 4), instead of the former five Groups 0, 1, 2, 3 and 4. Petrol and their mixtures with bioethanol belong to Group 1, while diesel and their mixtures with biodiesel belong to Group 4.

This Standard specifies an assay method, utilizing either manual or automated equipment. However, its latest edition sets the automated alternative as the reference method in the event of a dispute, unless otherwise agreed.

The latest edition of the Standard includes a specific point related to the test procedure utilizing automated equipment. Although the former edition allowed the use of this type of equipment, the procedure had not been defined so far. Nowadays, most distillation tests are carried out utilizing automated equipment. Once programmed, the procedure is performed in an autonomous way by the instrument, but programming them is not easy. It requires an absolute control and knowledge of the process as well as of the conditions required for each specific sample. The validity on the findings remains dependant on performing a validation study.

The latest edition of the Standard requires the establishment of new criteria for calculating repeatability and reproducibility.

The latest Standard specifies the percentages of distillate required at specific temperatures for petrol. Accordingly, the Standard includes a normative annex (Annex C) which provides values of reproducibility of petrol percent volume evaporated at 70, 100, 150 and 180°C. These reproducibility statements were estimated for the specification temperatures and percentages from the data collected in an inter-laboratory study:

- E70 → R = 2.7% V/V.
- E100 → R = 2.2% V/V.
- E150 → R = 1.3% V/V.
- E180 → R = 1.1% V/V.

4.2. Biofuel impact on fuel volatility

4.2.1. Petrol

As we set before, the petrol distillation curve study was developed through its analysis in three different sections. From the study of the two first sections (IBP-E20; E20-E70), we conclude that the evaporated percentage is higher in blends. So, in the range between 30 and 90°C, which comprises from IBP to the middle boiling point, the petrol volatility increases from bioethanol addition. In the final section, the sample distillation curves coincide to a great extent, and there are no any significant changes, as might be expected from the bioethanol boiling point.

4.2.2. Diesel

For diesel blends, containing up to 7% V/V biodiesel, the study of the distillation curve indicates the recovered percentage has risen slightly in the first section, while we can observe a bigger variation in the last one, where the fractioning of the heavy components of the samples takes place; this is between R90 and R95.

Diesel blends, containing 30% V/V biodiesel, are outside the ISO 3405:2011 Standard scope. Distillation test has been carried out, considering them as part of Group 4 and, then, accordingly to the conditions established for that group.

Distillation curve for diesel blends, containing 30% V/V biodiesel, indicates clearly the volatility diminution from the own characteristics of the biodiesel: high boiling point and low volatility (**Figure 15**). As in this type of products, the volatility diminution is directly related to the viscosity; the use of this type of blends in the engines is intimately related to their operation and performance.

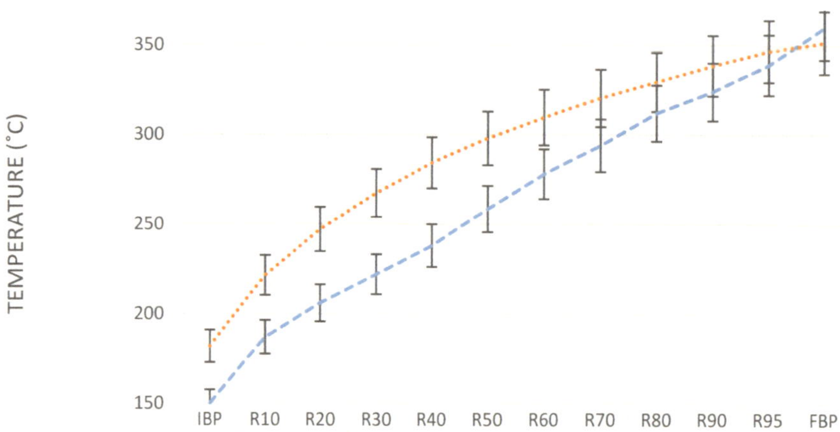

	IBP	R10	R20	R30	R40	R50	R60	R70	R80	R90	R95	FBP
0,0% FAME	150	187	206	222	238	259	278	294	312	324	339	360
30,0% FAME	181,9	221,4	247,1	267,3	284,2	297,9	309,6	320,4	329,4	338,6	346,6	351,5

Figure 15. Diesel distillation curves, 0.0% V/V and 30.0% V/V FAME.

4.3. Estimation of variation suffered by the accuracy of the automated method based on criteria of repeatability and reproducibility.

The latest edition of the Standard, ISO 3405:2011, establishes new repeatability and reproducibility criteria. This new criteria has reduced measure variations allowed in the whole range, affecting petrol (Group 1) and diesel (Group 4). Relatively small variations are allowed in the middle section of the distillation curve, whereas a more significant variation is allowed in the IBP and FBP determination, as shown in **Figures 16** and **17**.

4.4. Future prospects

The EU has promoted turning plants into fuel as a way to reduce carbon emissions from transport for the last 10 years. Today however, some say that biofuels have become part of the problem and actually have generated more CO_2 than they saved, as the demand for crops needed to produce them has led to the destruction of forests. The EU now wants to limit the amount of fuel produced from food crops and shift to biofuels that are produced from non-food sources, such as waste.

QUALITY CONTROL - CRM GASOLINE								
CRM			ISO 3405:2011			ISO 3405:2000		
CERTIFICATE VALUE (°C)	TOLERANCE RANGE CERTIFICATE (°C)	EVAPORATED %V/V	TOLERANCE RANGE SIMPLE	REPEATABILITY	REPRODUCIBILITY	TOLERANCE RANGE SIMPLE	REPEATABILITY	REPRODUCIBILITY
32,2	3,5	PI	3,54	2,46	5,00	5,09	3,90	7,20
49,3	2,3	10%	2,26	1,33	3,20	4,67	2,79	6,60
83,7	1,3	50%	1,33	0,74	1,88	3,08	1,72	4,35
143,9	2,7	90%	2,74	1,54	3,87	2,64	2,15	3,73
166,3	4,8	PF	4,79	3,33	6,78	6,29	4,40	8,90

Figure 16. The values of repeatability and reproducibility for the same petrol-CRM, according to ISO 3405:2011 and ISO 3405:2000.

QUALITY CONTROL - CRM DIESEL								
PAC-1037			ISO 3405:2011			ISO 3405:2000		
CERTIFICATE VALUE (°C)	TOLERANCE RANGE CERTIFICATE (°C)	EVAPORATED %V/V	TOLERANCE RANGE SIMPLE	REPEATABILITY	REPRODUCIBILITY	TOLERANCE RANGE SIMPLE	REPEATABILITY	REPRODUCIBILITY
172,90	6,00	PI	6,72	3,11	9,51	6,01	3,50	8,50
209,20	5,80	10%	3,25	1,97	4,60	5,49	3,76	7,76
274,20	6,10	50%	2,10	0,94	2,97	6,19	3,30	8,76
339,20	4,80	90%	3,60	1,39	5,09	4,93	3,22	6,97
354,90	7,00	95%	6,42	3,26	9,08	6,34	4,08	8,97
364,20	7,40	PF	5,02	2,20	7,10	7,42	3,50	10,50

Figure 17. The values of repeatability and reproducibility for the same diesel-CRM, according to ISO 3405:2011 and ISO 3405:2000.

The EU is committed to meeting 10% of its transport fuel needs from renewable sources, mostly biofuels, by 2020. But pressure is growing to change this policy by limiting the amount of food-based biofuels.

In response to these concerns, the European Parliament voted in favour of a compromise deal with the Council to limit the amount of fuel produced from food crops such as rapeseed and palm oil. The new legislation will enter into force in 2017 [12].

The future liquid automotive fuel formulations will remain dependant on distillation test as a powerful quality control tool. It provides useful information, which can be employed to establish optimal conditions of use in the appropriate engines, as well as to establish safe conditions for storage, transportation and delivery.

5. Conclusion

In 2011, the addition of up to 10% V/V bioethanol in petrol and up to 20% V/V of bio-diesel in diesel, as a result of biofuel promotion, has prompted a careful review of the

ISO 3405 Standard: Petroleum Products. Determination of Distillation Characteristics at Atmospheric Pressure.

Changes include:

1. The broadening of the scope to cover these new products
2. The adjustment of the calculations related to the method accuracy

Standard relating to test standardization, which establishes the procedure to determine the distillation characteristics of petroleum products at atmospheric pressure, remains open to future review and updating.

Author details

Mª Mercedes del Coro Fernández-Feal*, Luis R. Sánchez-Fernández and Blanca Sánchez-Fernández

*Address all correspondence to: coro.ffeal@udc.es

Fuels Laboratory, University of A Coruña, Spain

References

[1] Satou, M., Itoh, D., Hattori, H., Yoshida, T. Evaluation of ring size distribution in a heavy oil based on boiling point and molecular weight distributions. Fuel. 2000; **79**(3-4):339-348. DOI: 10.1016/S0016-2361(99)00168-4.

[2] Lluch Urpí, J. Tecnología y margen del refino de petróleo (Technology and petroleum refining profit). España: Ediciones Díaz de Santos; 2008. 464 pp. ISBN: 9788479788759.

[3] Álvarez Flórez, J.A., Callejón Agramunt, I., Forns Farrús, S. Motores Alternativos de Combustión Interna (Alternative Internal Combustion Engines). España: Ediciones UPC; 2005. 513 pp. ISBN. 9788483018187.

[4] Wilfrid, F. Los combustibles y su Tecnología (Fuels and their Technology). Bilbao, España: Ediciones Urmo S.A.; 1969. 688 pp. ISBN: 9788431400095.

[5] Sánchez-Fernández, B., Bouza-Fernández, S., Sánchez-Fernández, L.R., Seoane-López, L.R., Seoane-López, S., Fernández-Feal, M.L., Fernández-Feal, M.C. Biodiésel en el Mercado Gallego (Objetivo 2010) como consecuencia del Plan de Fomento de Energías Renovables (Biodiesel in the Galician Market (Objective 2010) as a result of the Plan for the Promotion of Renewable Energies). In: Proceedings World Economy Meeting XII; Santiago de Compostela, España; 2010. ISBN 9788415026068.

[6] Aburudyna, A., Karonis, D., Zannikos, F., Lois, E.. Impact of biodiesel addition on distillation characteristics and cetane number of diesel fuels. Journal of Energy Engineering. 2015; 141(2). DOI: 10.1061(ASCE) EY.1943-7897.0000244.

[7] Daganzo, J.M. Blog de Innovación Tecnológica (REPSOL). [Internet]. 05.01.2012 [Updated: 22.07.2016].Available from: http://blogs.repsol.com/innovacion/combustibles-convencionales-y-biocombustibles-un-ejemplo-de-convivencia [Accessed: 20.05.2016].

[8] EN 228:2012. Automotive fuels. Unleaded petrol. Requirements and test methods.

[9] EN 590:2013. Automotive fuels. Diesel. Requirements and test methods.

[10] ISO 3405:2000. Petroleum products. Determination of distillation characteristics at atmospheric pressure.

[11] ISO 3405:2011. Petroleum products. Determination of distillation characteristics at atmospheric pressure.

[12] Parlamento Europeo. En portada. Punto de mira. [Internet]. 14.04.2015 [Updated: 05.05.2016]. Available from: http://www.europarl.europa.eu/news/es/news-room/20130719STO17435/Nueva-pol%C3%ADtica-de-biocombustibles-menos-CO2-y-m%C3%A1s-seguridad-alimentaria [Accessed: 15.07.2016].

Chapter 5

Enhanced Distillation Under Infrared Characteristic Radiation

Kuo-Ting Wang, M. Quinn Brewster and Wei-Hsiang Lai

Additional information is available at the end of the chapter

http://dx.doi.org/10.5772/67401

Abstract

This chapter introduces quasi-steady water vaporization under mid-infrared (IR) radiation and the IR absorption of characteristic radiation associated with the first-kind liquid-gaseous phase transition of water. When characteristic radiation in the mid-IR spectral range is applied to water surface, the strong volumetric absorption of radiation energy in the liquid-phase causes water to be nearly isothermal. In addition to volumetric absorption, surface absorption of characteristic radiation induces vaporization of water. The complete mechanism of liquid-gaseous phase-transition radiation involves the direct surface absorption/emission of infrared energy accompanied by evaporation/condensation of water. A direct consequence of excess characteristic radiation upon water surface is the induced supersaturation. This mechanism opens up a door for enhanced distillation under characteristic radiation. Blackbody-like materials such as black anodized aluminum surfaces and metal surfaces painted in black are recommended to be heated to ~250°C to serve as economical radiation sources. For isothermal water at room temperatures, ~20% supersaturation can be induced by hemispherical Blackbody radiation with temperature ~11°C higher than the water temperature. In this situation, energy extracted from the ambient for water vaporization can be as much as 80% of latent heat. With radiation-enhanced evaporation, the production cost for distilled water is significantly reduced as compared to distillation at the boiling point.

Keywords: water, evaporation, supersaturation, infrared, radiation, phase-transition radiation

1. Introduction

While latent heat of evaporation is a fixed amount of energy needed to vaporize liquid-water, how energy is supplied to cause evaporation is an open area for engineers to explore to achieve their design goals. Evaporation is a process that water molecules go through the phase transition of the first kind from the liquid phase to the gaseous phase. This process is driven by the concentration gradient of water molecules on the vapor side of the water surface. When water is heated up to the boiling point, the saturation pressure of water is increased to exceed the ambient water-vapor pressure not only on the water surface but also in the bulk so that "evaporation" (*boiling*) takes place everywhere in the water. For water purification purposes, water can be distilled in this way. One advantage of this type of water purification is that distillation can be quickly carried out by supplying sufficient heat. However, one obvious disadvantage at the same time is that a significant amount of energy is lost to the ambient in order to maintain water temperature at the boiling point.

This chapter suggests another way of distillation that enables water to be supersaturated on the surface so as to enhance evaporation. This method utilizes the mechanism of so-called phase-transition radiation to enhance evaporation without requiring water to reach the boiling point. Since liquid-water molecules are constantly experiencing breaking and formation of hydrogen bonds due to intermolecular vibrations, there are moments when some water molecules are weakly bonded. When photons hit such molecules on the surface, the photons that carry enough energy to break the hydrogen bond will be absorbed to cause transitions of energy states, namely evaporation. The spectral range of the evaporative absorption defines the spectrum of characteristic radiation, which is in the mid-IR range. A good amount of effort [1] has been spent on the characteristic wavelength of phase-transition radiation in vapor condensation process.

Quasi-steady equations for radiation absorption in semi-infinite liquid-water as well as their dilute approximations are presented in this chapter. Semi-infinite liquid-water is selected to represent a common scenario for engineering distillation applications. These equations deal with liquid phase, vapor phase, and the vapor liquid interface in the presence of IR characteristic radiation. In addition to volumetric absorption of radiation, which is commonly recognized as far as radiative heat transfer is concerned, surface absorption of radiation is taken into account in the liquid-phase equations. Based on the IR absorption characteristics in the liquid-water, the liquid phase is essentially isothermal under moderate radiation strength in the mid-IR range. In the equations for vapor phase and interface, the situation of supersaturation triggered by an excess amount of IR characteristic radiation is presented for the first time. Enhanced evaporation as a result of supersaturation is exemplified in dilute systems. Although the enhanced evaporation rate by IR radiation may not be as fast as that of the traditional way of boiling water, this method is more energy efficient and economical when continual distillation is needed for a larger amount of water.

In the last section of this chapter, the theory of phase-transition radiation is extended to the situation of supersaturation for semi-infinite water. The evaporation flux due to phase-transition radiation is linked to characteristic radiation and supersaturation. The equation for quasi-steady supersaturation is obtained and used to close the mathematic problem when vapor pressure at water surface is no longer subject to saturation conditions.

2. Quasi-steady semi-infinite water

2.1. Liquid phase

A sketch for the thermodynamic system of semi-infinite water is shown in **Figure 1**. In the liquid phase, the temperature distribution is subject to three heat transfer modes: conduction, convection, and radiation,

$$q = -k_l \frac{dT}{dx} - \dot{m}^{''} C(T + T_{ref}) + (1 - f_s) q_r \tag{1}$$

where q denotes heat flux, k_l thermal conductivity of liquid-water, $\dot{m}^{''}$ mass flux, C heat capacity of water, f_s the fraction of surface absorption of radiation, and q_r radiative heat flux. Surface absorption efficiency f_s is defined later in Section 2.4 and its values ($0 \leq f_s \leq 1$) are plotted in **Figure 6**.

A general temperature equation can be obtained by imposing constant heat flux conditions at quasi-steady state with respect to water depth (i.e., x-direction in **Figure 1**),

$$\frac{dq}{dx} = 0 \Rightarrow k_l \frac{d^2T}{dx^2} + \dot{m}^{''} C \frac{dT}{dx} - \left(1 - f_s\right) \frac{dq_r}{dx} = 0 \tag{2}$$

For the radiative heat transfer with absorption coefficient K_a in liquid-water,

$$\frac{dq_r}{dx} = -K_a q_r \Rightarrow q_r = q_{r,s} e^{-K_a x} \Rightarrow \frac{dq_r}{dx} = -q_{r,s} K_a e^{-K_a x}, \tag{3}$$

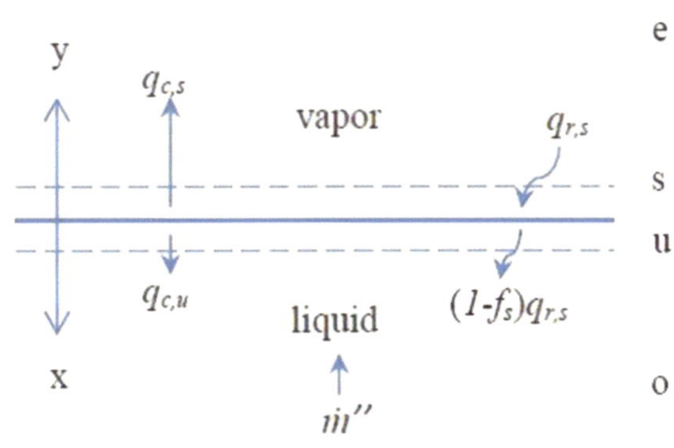

Figure 1. Definitions for thermodynamic states, energy fluxes, and coordinate systems at vapor-liquid interface for semi-infinite water.

which leads to

$$k_l \frac{d^2T}{dx^2} + \dot{m}''C\frac{dT}{dx} + (1-f_s)q_{r,s}K_a e^{-K_a x} = 0. \tag{4}$$

Temperature boundary conditions (BCs) at surface (i.e., T_s at $x = 0$) and at a sufficiently deep location (either T at $x \rightarrow \infty$ or $dT/dx = 0$ at $x \rightarrow \infty$) can be specified to solve Eq. (4).

BCs: $T|_{x=0} = T_s$, $\frac{dT}{dx}\big|_{x\rightarrow\infty} = 0$ or $T|_{x\rightarrow\infty} = T_o$.

2.1.1. Nondimensionalization

For small temperature changes (i.e., a few degrees C), physical properties (K_a, k_l, C, f_s) can be assumed to be constant and a general form for the liquid-phase equation is obtained following a nondimensionalization analysis with,

$$\delta_r = 1/K_a, \qquad \delta_t = k_l/(\dot{m}''C), \qquad \overline{x} = x/\delta_r,$$
$$\Delta T_{\max} = (1-f_s)q_{r,s}/\left[k_l(\delta_r^{-1} - \delta_t^{-1})\right], \qquad \overline{T} = (T - T_s)/\Delta T_{\max}, \tag{5}$$
$$\frac{d^2\overline{T}}{d\overline{x}^2} + \frac{\delta_r}{\delta_t}\frac{d\overline{T}}{d\overline{x}} + \left(1 - \frac{\delta_r}{\delta_t}\right)e^{-\overline{x}} = 0.$$

In Eq. (5) different scales for length (characteristic length for IR absorption δ_r and convective length scale δ_t) as well as temperature (ΔT_{max}) [2] are used to formulate the nondimensional form. Based on two different kinds of BCs, two different forms of solutions are obtained,

General solution: $\overline{T} = A + Be^{-\overline{x}(\delta_r/\delta_t)} - e^{-\overline{x}}$ (A, B: constants);

Case 1: BCs $\overline{T}|_{\overline{x}=0} = 0$, $\frac{d\overline{T}}{d\overline{x}}\big|_{\overline{x}\rightarrow\infty} = 0$, solution: $\overline{T} = e^{-\overline{x}(\delta_r/\delta_t)} - e^{-\overline{x}}$;

Case 2: BCs $\overline{T}|_{\overline{x}=0} = 0$, $\overline{T}|_{\overline{x}\rightarrow\infty} = (T_o - T_s)/\Delta T_{\max} = \overline{T_o}$, solution: $\overline{T} = \overline{T_o} + (1 - \overline{T_o})e^{-\overline{x}(\delta_r/\delta_t)} - e^{-\overline{x}}$.

2.1.2. Isothermal water under mid-IR

Based on the mid-IR absorption spectrum of liquid-water, water is fairly opaque under mid-IR radiation and the attenuation of mid-IR radiation occurs within only several microns [2]. The characteristic length for IR absorption δ_r (which is inversely proportional to the absorption coefficient) is much smaller than the convective length scale of water δ_t in typical conditions. As a result, the convection term in the liquid-phase temperature equation can be ignored. The simplified equation has the following form,

$$\frac{\delta_r}{\delta_t} \approx 0 \Rightarrow \frac{d^2\overline{T}}{d\overline{x}^2} + e^{-\overline{x}} \approx 0, \quad \Delta T_{\max} \approx (1-f_s)q_{r,s}\delta_r/k_l, \tag{6}$$

and the same simplified solution for both Cases 1 and 2,

http://dx.doi.org/10.5772/67401

$$\overline{T} \approx 1 - e^{-\overline{x}} \text{ or } T \approx T_s + \Delta T_{\max}\left(1 - e^{-K_a x}\right). \tag{7}$$

Eq. (7) leads to an important feature for liquid-phase quasi-steady temperature distribution: water is essentially isothermal under mid-IR radiation with moderate strength [2], as illustrated in Example 1.

The liquid-phase enthalpy changes from the top ($x = 0$) to the bottom ($x \to \infty$) can be described in terms of conductive and radiative heat transfer by applying integration to Eq. (4) from $x = 0$ to $x \to \infty$:

$$k_l \frac{dT}{dx}\bigg|_{x=0}^{x\to\infty} + \dot{m}''CT\big|_{x=0}^{x\to\infty} - \left(1 - f_s\right)q_{r,s}e^{-K_a x}\big|_{x=0}^{x\to\infty} = 0 \Rightarrow \left(q_{c,u} - q_{c,o}\right)$$
$$+ \dot{m}''\left(h_{1,o} - h_{1,u}\right) + \left(1 - f_s\right)q_{r,s} = 0.$$

Normally the temperature gradient vanishes at the bottom, $q_{c,o} = 0$, giving,

$$\dot{m}''\left(h_{1,o} - h_{1,u}\right) = -\left(1 - f_s\right)q_{r,s} - q_{c,u}. \tag{8}$$

2.2. Vapor phase

Basic definitions related to water vapor-air mixtures in the vapor phase are reviewed before introducing vapor-phase equations. We shall limit our discussions to two-species mixtures.

2.2.1. Basic definitions

$$h = m_1 h_1 + m_2 h_2 \qquad \text{(mixture enthalpy)}$$

$$C_p = m_1 C_{p,1} + m_2 C_{p,2} \qquad \text{(mixture } C_p\text{)}$$

$$\rho = \rho_1 + \rho_2 \qquad \text{(mixture density)}$$

$$m_{1,2} = \frac{\rho_{1,2}}{\rho} \qquad \text{(mass fraction of species 1, vapor or 2, air)}$$

Assuming ideal gas for water vapor and air, vapor mass fraction m_1 becomes,

$$m_1 = \frac{\rho_1}{\rho} = \frac{P_1 M_1}{P_1 M_1 + P_2 M_2} = \frac{P_1}{\frac{M_2}{M_1}P + \left(1 - \frac{M_2}{M_1}\right)P_1} = \frac{(RH)P_{sat}}{1.61P - 0.61(RH)P_{sat}}, \tag{9}$$

where the ratio of air molecular weight to vapor, $M_2/M_1 = 28.97/18 = 1.61$, and the partial pressure of vapor $P_1 = (RH)\,(P_{sat})$. This form of m_1 and its dilute approximation are listed in **Table 3**. Reference values for air molecular weight and other properties are available at [3].

2.2.2. *Mass transfer of water vapor*

The transport of water vapor can be divided into two parts: the microscopic molecular diffusion and the macroscopic convection. When vapor mass fraction is much less than unity, i.e., $m_1 << 1$, the macroscopic convection term is often negligible, leaving the diffusion term described by Fick's law in the dilute system. Mass flux of water vapor is described by,

$$\dot{m}_1'' = \underbrace{m_1 \rho v}_{convection} \underbrace{- \rho D \frac{dm_1}{dy}}_{diffusion}. \tag{10}$$

The quasi-steady equation for vapor mass transport is obtained assuming constant D,

$$\frac{d\left(\dot{m}_1''\right)}{dy} = 0 \Rightarrow \rho D \frac{d^2 m_1}{dy^2} = \rho v \frac{dm_1}{dy}. \tag{11}$$

The algebraic formula for mass diffusivity D is listed in **Table 1**.

Output	**Input**	**Algebraic formula**	**Max error**	**Reference**
T_{sat} [°C]	P_{sat} [atm]	$T_{sat} = \frac{237.3 \times \log_{10}\left(\frac{P_{sat}}{0.00603}\right)}{7.5 - \log_{10}\left(\frac{P_{sat}}{0.00603}\right)}$	<1%	[4]
P_{sat} [atm]	T_{sat} [°C]	$P_{sat} = 0.00603 \times 10^{\frac{7.5T_{sat}}{237.3+T_{sat}}}$	<1%	[4]
D [m²/s]	P [atm], T [°C]	$D = \left(1.97 \times 10^{-5}\right)\left(\frac{273.15+T}{256}\right)^{1.685}\left(\frac{1.013}{P}\right)$		[2]
α_{sat} [m²/s]	T [°C]	$\alpha_{sat} = \sum_{i=0}^{4} SA_i T_i^{\,i}$		[5]
α_{dry} [m²/s]	k [W/m-K], ρ [g/m³], C_p [kJ/kg-K]	$\alpha_{dry} = \frac{k}{\rho C_p}$		Definition
k [W/m-K]	T [K]	$k = \sum_{i=0}^{5} KA_i T_i^{\,i}$		[5]
$C_{p,2}$ [kJ/kg-K]	T [K]	$C_{p,2} = \sum_{i=0}^{4} CA_i T_i^{\,i} \approx 1$		[5]
ρ_2 [g/m³]	P [Pa], T [K]	$\rho_2 = \frac{PM_{air}}{RT}$		Ideal gas
h_2 [kJ/kg]	T [°C]	$h_2 = 2501 + T$		
$h_{1,u}$ [kJ/kg]	T [°C]	$h_{1,u} = 4.2T$	<1%	

Errors are based on comparisons with available data from steam tables for T = 1–100°C.

Table 1. Algebraic formulae for selected properties.

2.2.3. *Heat transfer in water vapor-air mixtures*

For the mixture heat flux, conduction is included in addition to diffusion and convection,

$$q = \underbrace{\rho v h}_{convection} \underbrace{- \rho D \sum_{i=1}^{2} h_i \frac{dm_i}{dy}}_{diffusion} \underbrace{- k \frac{dT}{dy}}_{conduction} = \left(\rho v h - \rho \frac{k}{\rho C_p} \frac{dh}{dy}\right) + \sum_{i=1}^{2} h_i \frac{dm_i}{dy}\left(\rho \frac{k}{\rho C_p} - \rho D\right). \tag{12}$$

In the two-species system, since $m = m_1 + m_2 = 1$ = constant, it follows that $dm_2 = -dm_1$, and,

$$q = \left(\rho v h - \rho\alpha\frac{dh}{dy}\right) - \left(\rho D\frac{dm_1}{dy}\right)(h_1 - h_2)\left(1 - \frac{1}{Le}\right), \tag{13}$$

where Lewis number $Le = D/\alpha$ and thermal diffusivity $\alpha = k/\rho C_p$. Numerical values of Le for dry air and saturated air at different temperatures are tabulated in **Table 4** and plotted in **Figure 2**. The algebraic formula for α can be found in **Table 1** with the corresponding curve-fit coefficients available in **Table 2** [5].

The heat flux equation can be simplified if any of the following conditions are valid: (1) dilute; (2) nearly equal enthalpy for two species; (3) $Le \approx 1$, which is generally true for air-vapor mixtures under atmospheric conditions. The condition of nearly equal enthalpy is achievable with properly chosen reference enthalpies for air and vapor. To achieve this condition, enthalpies of liquid-water and vapor are taken from steam tables and the reference enthalpy of air is set to $h_{2,ref}$ = 2501 kJ/kg at T_{ref} = 0°C to give,

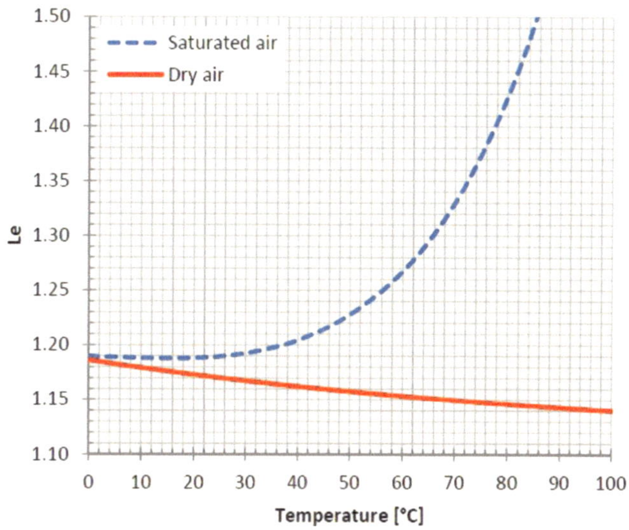

Figure 2. Lewis number, *Le*, for saturated air and dry air at one atmosphere. Numerical values are shown in **Table 4**.

α_{sat} [m²/s]	k [W/m-K]	$C_{p,2}$ [kJ/kg-K]	ρ [g/m³]
SA_0 = 1.847185729E-5	KA_0 = −2.276501E-3	CA_0 = 1.03409	M_{air} = 28.97 [g/mol] [3]
SA_1 = 1.161914598E-7	KA_1 = 1.2598485E-4	CA_1 = −2.84887E-4	R = 8.314 [J/mol-K]
SA_2 = 2.373056947E-10	KA_2 = −1.4815235E-7	CA_2 = 7.816818E-7	
SA_3 = −5.769352751E-12	KA_3 = 1.73550646E-10	CA_3 = −4.970786E-10	
SA_4 = −6.369279936E-14	KA_4 = −1.066657E-13	CA_4 = 1.077024E-13	
	KA_5 = 2.47663035E-17		

Table 2. A supplemental table with numerical coefficients and constants for **Table 1**.

Variable	Original form	Algebraic formula	Dilute approximation (m_1<<1)
m_1	$\frac{\rho_1}{\rho}$	$\frac{(RH)P_{sat}}{1.61P-0.61(RH)P_{sat}}$	$\frac{(RH)P_{sat}}{1.61P}$
$\dot{m}''$	$\frac{\rho D}{L}\ln(1+B_m)$	$\frac{\rho D}{L}\ln\left(1+\frac{m_{1,s}-m_{1,e}}{1-m_{1,s}}\right)$	$\rho D\frac{m_{1,s}-m_{1,e}}{L}$; $\frac{\rho D}{L}\frac{(RH)_s P_{sat,s}-(RH)_e P_{sat,e}}{1.61P}$
$\dot{m}''$	$\frac{\rho\alpha}{L}\ln(1+B_h)$	$\frac{\rho\alpha}{L}\ln\left[1+\frac{h_e-h_s}{h_s-\left(h_{1,o}+q_{r,s}/\dot{m}''\right)}\right]$	$\frac{\rho\alpha}{L}\frac{C_p\left(T_e-T_s+T_{r,s}\right)}{\left(h_s-h_{1,o}\right)}$, $T_{r,s}=\frac{q_{r,s}L}{k}$
C_p	$m_2C_{p,1}+m_2C_{p,2}$		$C_{p,2}$
H	$m_1h_1+m_2h_2$		h_2

Table 3. Different forms of selected variables and their dilute approximations.

$$h_2[\text{kJ/kg}] = h_{2,ref} + C_{p,2}T\ [^\circ\text{C}] \approx 2501 + T\ [^\circ\text{C}], \tag{14}$$

where $C_{p,2} \approx 1$ kJ/kg-K for T = 0–100°C. The liquid-water enthalpy is approximated (within 1% errors for T = 1–99°C as compared to steam tables) by,

$$h_{1,u}[\text{kJ/kg}] \approx 4.2T\ [^\circ\text{C}]. \tag{15}$$

This leads to a simplified heat flux equation,

$$q \approx \left(\rho v h - \rho\alpha\frac{dh}{dy}\right), \tag{16}$$

and its quasi-steady equation,

$$\frac{dq}{dy} = 0 \Rightarrow \rho\alpha\frac{d^2h}{dy^2} \approx \rho v\frac{dh}{dy}. \tag{17}$$

Dilute approximations for mixture enthalpy h and specific heat C_p are listed in **Table 3**. Algebraic forms for air enthalpy h_2 and specific heat $C_{p,2}$ are tabulated in **Table 1**. Curve-fit coefficients for $C_{p,2}$ are placed in **Table 2** [5].

2.3. Vapor-liquid interface

Vapor-liquid interface equations are obtained by matching the mass flux of species 1 (vapor or liquid-water) and energy flux at the interface.

Mass flux of species 1:

$$\overbrace{\dot{m}'' = \dot{m}_u''}^{liquid} = \dot{m}_{1,u}'' = \overbrace{\dot{m}_{1,s}'' = \underbrace{m_{1,s}\rho v}_{convection} - \underbrace{\rho D\left.\frac{dm_1}{dy}\right|_s}_{diffusion} = m_{1,s}\dot{m}'' - \rho D\left.\frac{dm_1}{dy}\right|_s}^{vapor}$$

$$\Rightarrow \dot{m}'' = -\frac{\rho D}{1-m_{1,s}}\left.\frac{dm_1}{dy}\right|_s. \tag{18}$$

Together with the mass diffusion equation (11), mass flux can be solved for,

$$\dot{m}'' = \frac{\rho D}{L}\ln(1+B_m), \quad B_m = \frac{m_{1,s} - m_{1,e}}{1 - m_{1,s}}. \tag{19}$$

Its dilute approximation is listed in **Table 3**.

Heat flux:

$$q = q_u = \overbrace{\underbrace{-q_{c,u}}_{conduction} + \underbrace{\dot{m}'' h_{1,u}}_{convection} + \underbrace{f_s q_{r,s}}_{radiation}}^{liquid} = \overbrace{q_s \approx \dot{m}'' h_s - \rho\alpha \left.\frac{dh}{dy}\right|_s}^{vapor}$$

$$\Rightarrow \dot{m}''\left[h_s - \left(h_{1,u} - \frac{q_{c,u}}{\dot{m}''} + \frac{f_s q_{r,s}}{\dot{m}''}\right)\right] \approx \rho\alpha \left.\frac{dh}{dy}\right|_s. \tag{20}$$

Eq. (20) is based on enthalpy formulation. Alternatively, the heat flux equation can be obtained based on temperature formulation without approximations,

$$q = q_u = \overbrace{\underbrace{-q_{c,u}}_{conduction} + \underbrace{\dot{m}'' h_{1,u}}_{convection} + \underbrace{f_s q_{r,s}}_{radiation}}^{liquid} = \overbrace{q_s = \dot{m}'' h_{1,s} - k \left.\frac{dT}{dy}\right|_s}^{vapor}$$

$$\Rightarrow \dot{m}'' h_{fg} = \dot{m}''\left(h_{1,s} - h_{1,u}\right) = -q_{c,u} + f_s q_{r,s} + k \left.\frac{dT}{dy}\right|_s. \tag{21}$$

Based on Eq. (20) and the heat diffusion equation (17), an alternative form for mass flux can be obtained to relate it to heat transfer variables,

$$\dot{m}'' \approx \frac{\rho\alpha}{L}\ln(1+B_h), \quad B_h = \frac{h_e - h_s}{h_s - \left(h_{1,u} - \frac{q_{c,u}}{\dot{m}''} + \frac{f_s q_{r,s}}{\dot{m}''}\right)}. \tag{22}$$

At the first glance evaporation flux in Eq. (22) seems to be influenced by f_s. However, its dependency on f_s is eliminated [2] when the energy balance for the entire liquid is taken into account as done in Example 2, giving,

$$B_h = \frac{h_e - h_s}{h_s - \left(h_{1,o} + q_{r,s}/\dot{m}''\right)}. \tag{23}$$

The corresponding dilute approximation for mass flux is shown in **Table 3**.

Given $(RH)_e$, these two forms of mass flux are instrumental in evaluating surface temperature T_s—even without the knowledge of the vapor layer thickness L. For an isothermal dilute system (see approximated forms in **Table 3**), an implicit form for T_s is obtained by equating the two mass-flux forms, giving,

$$\frac{(RH)_s P_{sat,s} - (RH)_e P_{sat,e}}{1.61P(T_e - T_s + T_{r,s})} \approx \frac{C_p}{Le} \frac{1}{\underbrace{2501 + T_s}_{h_s} - \underbrace{4.2T_s}_{h_{1,u}}}, \tag{24}$$

or,

$$\frac{P_{sat,s}}{P} \approx \frac{P_{sat,e}}{P} \frac{(RH)_e}{(RH)_s} + \frac{1.61C_p}{(RH)_s Le} \frac{(T_e - T_s + T_{r,s})}{2501 - 3.2T_s}, \tag{25}$$

in which the unit for temperature (i.e., for T_e, T_s, and $T_{r,s}$) is [°C]. **Figure 2** shows the relationship of Lewis number Le with T and RH at one atmosphere. For evaporation taking place at a moderate-temperature (0–40°C) under atmospheric conditions, $Le \approx 1.18$ appears to be a good approximation for water vapor-air mixtures.

To facilitate numerical calculations, algebraic formulas for thermal properties are listed below.

2.4. Supersaturation by characteristic radiation

2.4.1. Energy distributions of water molecules in gaseous and liquid phases

Unlike conventional two-energy-level models characterized by discrete absorption lines, phase-transition radiation is accompanied by a continuous absorption band. The "band" feature of the characteristic radiation is primarily a result of the convolution of population distributions for translational energy and rotational energy of vapor molecules. Since vapor molecules can be treated as a continuum in atmospheric conditions, the population distribution of translational energy at thermodynamic equilibrium is described by the Maxwell-Boltzmann distribution. Consequently vapor molecules are continuously populated as demonstrated in **Figure 3**. Descriptions for this model can be found in [1]. Intermolecular vibrations of water molecules in the liquid phase broaden the characteristic radiation band and smooth out its far wings.

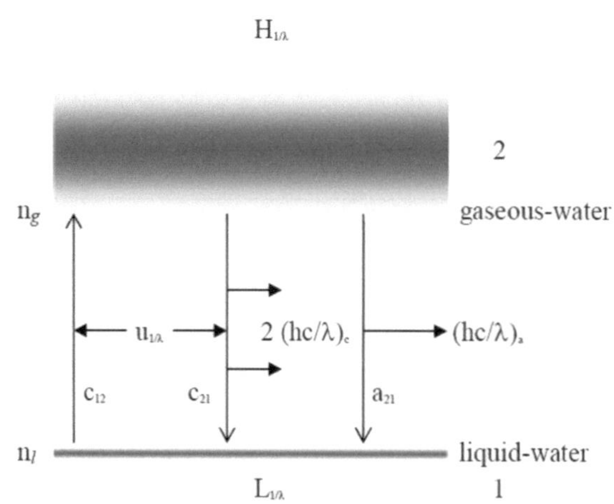

Figure 3. Energy states (2: gaseous; 1: liquid) and phase-transition radiation in the condensation/evaporation process.

http://dx.doi.org/10.5772/67401

Although the peak value of the population distribution function is slightly brought down as a result of the broadening at the lower energy state, a previous work [6] has shown that the shapes of population distributions with and without invoking broadening effects of intermolecular vibrations in liquid-water are similar to each other.

For modeling purposes, vapor molecules are distributed over a continuous band by energy convolution and liquid-water molecules are treated as at the same energy level to simplify mathematical formulations. In other words, the broadening of the lower energy state is ignored. Modeling intermolecular vibrations of water molecules is in fact a very challenging task because the hydrogen bonds, which directly affect intermolecular vibrations, are constantly breaking and forming and are highly dependent on temperature.

Since mathematical forms for population distribution functions for vapor molecules have been established in the previous work [6], their derivations are skipped in this chapter. Together with population distribution functions and related parameters, equations for evaporation fluxes related to absorption, spontaneous emission, and induced emission are outlined in **Table 5**.

T [°C]	Dry air, α_{dry} [m^2/s]	Saturated air, α_{sat} [m^2/s]	D [m^2/s]	Dry air, Le	Saturated air, Le
0	1.853E-05	1.847E-05	2.197E-05	1.186	1.190
1	1.865E-05	1.859E-05	2.211E-05	1.185	1.189
2	1.878E-05	1.871E-05	2.225E-05	1.185	1.189
3	1.891E-05	1.882E-05	2.238E-05	1.184	1.189
4	1.903E-05	1.894E-05	2.252E-05	1.183	1.189
5	1.916E-05	1.906E-05	2.266E-05	1.183	1.189
6	1.929E-05	1.918E-05	2.279E-05	1.182	1.189
7	1.941E-05	1.929E-05	2.293E-05	1.181	1.188
8	1.954E-05	1.941E-05	2.307E-05	1.181	1.188
9	1.967E-05	1.953E-05	2.321E-05	1.180	1.188
10	1.980E-05	1.965E-05	2.335E-05	1.179	1.188
11	1.993E-05	1.977E-05	2.349E-05	1.179	1.188
12	2.006E-05	1.989E-05	2.363E-05	1.178	1.188
13	2.019E-05	2.001E-05	2.377E-05	1.177	1.188
14	2.032E-05	2.013E-05	2.391E-05	1.177	1.188
15	2.045E-05	2.025E-05	2.405E-05	1.176	1.188
16	2.058E-05	2.036E-05	2.419E-05	1.175	1.188
17	2.071E-05	2.048E-05	2.433E-05	1.175	1.188
18	2.084E-05	2.060E-05	2.447E-05	1.174	1.188
19	2.097E-05	2.072E-05	2.461E-05	1.174	1.188
20	2.110E-05	2.083E-05	2.475E-05	1.173	1.188

T [°C]	Dry air, α_{dry} [m^2/s]	Saturated air, α_{sat} [m^2/s]	D [m^2/s]	Dry air, Le	Saturated air, Le
21	2.123E-05	2.095E-05	2.490E-05	1.172	1.188
22	2.137E-05	2.107E-05	2.504E-05	1.172	1.189
23	2.150E-05	2.118E-05	2.518E-05	1.171	1.189
24	2.163E-05	2.130E-05	2.532E-05	1.171	1.189
25	2.177E-05	2.141E-05	2.547E-05	1.170	1.190
26	2.190E-05	2.152E-05	2.561E-05	1.170	1.190
27	2.203E-05	2.163E-05	2.576E-05	1.169	1.191
28	2.217E-05	2.175E-05	2.590E-05	1.168	1.191
29	2.230E-05	2.186E-05	2.605E-05	1.168	1.192
30	2.244E-05	2.196E-05	2.619E-05	1.167	1.193
31	2.257E-05	2.207E-05	2.634E-05	1.167	1.193
32	2.271E-05	2.218E-05	2.648E-05	1.166	1.194
33	2.284E-05	2.228E-05	2.663E-05	1.166	1.195
34	2.298E-05	2.238E-05	2.678E-05	1.165	1.196
35	2.312E-05	2.249E-05	2.692E-05	1.165	1.197
36	2.325E-05	2.259E-05	2.707E-05	1.164	1.199
37	2.339E-05	2.268E-05	2.722E-05	1.164	1.200
38	2.353E-05	2.278E-05	2.737E-05	1.163	1.201
39	2.366E-05	2.287E-05	2.752E-05	1.163	1.203
40	2.380E-05	2.297E-05	2.766E-05	1.162	1.205
41	2.394E-05	2.306E-05	2.781E-05	1.162	1.206
42	2.408E-05	2.314E-05	2.796E-05	1.161	1.208
43	2.422E-05	2.323E-05	2.811E-05	1.161	1.210
44	2.436E-05	2.331E-05	2.826E-05	1.160	1.212
45	2.450E-05	2.339E-05	2.841E-05	1.160	1.215
46	2.464E-05	2.347E-05	2.856E-05	1.159	1.217
47	2.478E-05	2.355E-05	2.871E-05	1.159	1.219
48	2.492E-05	2.362E-05	2.887E-05	1.159	1.222
49	2.506E-05	2.369E-05	2.902E-05	1.158	1.225
50	2.520E-05	2.376E-05	2.917E-05	1.158	1.228
51	2.534E-05	2.382E-05	2.932E-05	1.157	1.231
52	2.548E-05	2.388E-05	2.947E-05	1.157	1.234
53	2.562E-05	2.394E-05	2.963E-05	1.156	1.238
54	2.576E-05	2.399E-05	2.978E-05	1.156	1.241
55	2.591E-05	2.404E-05	2.993E-05	1.156	1.245
56	2.605E-05	2.408E-05	3.009E-05	1.155	1.249

http://dx.doi.org/10.5772/67401

T [°C]	Dry air, α_{dry} [m^2/s]	Saturated air, α_{sat} [m^2/s]	*D* [m^2/s]	Dry air, *Le*	Saturated air, *Le*
57	2.619E-05	2.412E-05	3.024E-05	1.155	1.254
58	2.633E-05	2.416E-05	3.040E-05	1.154	1.258
59	2.648E-05	2.420E-05	3.055E-05	1.154	1.263
60	2.662E-05	2.423E-05	3.071E-05	1.153	1.268
61	2.677E-05	2.425E-05	3.086E-05	1.153	1.273
62	2.691E-05	2.427E-05	3.102E-05	1.153	1.278
63	2.705E-05	2.429E-05	3.117E-05	1.152	1.284
64	2.720E-05	2.430E-05	3.133E-05	1.152	1.289
65	2.734E-05	2.431E-05	3.149E-05	1.152	1.295
66	2.749E-05	2.431E-05	3.164E-05	1.151	1.302
67	2.764E-05	2.430E-05	3.180E-05	1.151	1.309
68	2.778E-05	2.429E-05	3.196E-05	1.150	1.316
69	2.793E-05	2.428E-05	3.212E-05	1.150	1.323
70	2.807E-05	2.426E-05	3.228E-05	1.150	1.330
71	2.822E-05	2.423E-05	3.243E-05	1.149	1.338
72	2.837E-05	2.420E-05	3.259E-05	1.149	1.347
73	2.852E-05	2.417E-05	3.275E-05	1.149	1.355
74	2.866E-05	2.412E-05	3.291E-05	1.148	1.364
75	2.881E-05	2.407E-05	3.307E-05	1.148	1.374
76	2.896E-05	2.402E-05	3.323E-05	1.148	1.384
77	2.911E-05	2.395E-05	3.339E-05	1.147	1.394
78	2.926E-05	2.388E-05	3.355E-05	1.147	1.405
79	2.941E-05	2.381E-05	3.371E-05	1.147	1.416
80	2.956E-05	2.372E-05	3.388E-05	1.146	1.428
81	2.970E-05	2.363E-05	3.404E-05	1.146	1.440
82	2.985E-05	2.353E-05	3.420E-05	1.146	1.453
83	3.000E-05	2.343E-05	3.436E-05	1.145	1.467
84	3.016E-05	2.332E-05	3.453E-05	1.145	1.481
85	3.031E-05	2.319E-05	3.469E-05	1.145	1.496
86	3.046E-05	2.307E-05	3.485E-05	1.144	1.511
87	3.061E-05	2.293E-05	3.502E-05	1.144	1.527
88	3.076E-05	2.278E-05	3.518E-05	1.144	1.544
89	3.091E-05	2.263E-05	3.534E-05	1.143	1.562
90	3.106E-05	2.247E-05	3.551E-05	1.143	1.580
91	3.122E-05	2.230E-05	3.567E-05	1.143	1.600
92	3.137E-05	2.211E-05	3.584E-05	1.143	1.621

T [°C]	Dry air, α_{dry} [m^2/s]	Saturated air, α_{sat} [m^2/s]	D [m^2/s]	Dry air, *Le*	Saturated air, *Le*
93	3.152E-05	2.192E-05	3.600E-05	1.142	1.642
94	3.167E-05	2.173E-05	3.617E-05	1.142	1.665
95	3.183E-05	2.152E-05	3.634E-05	1.142	1.689
96	3.198E-05	2.130E-05	3.650E-05	1.141	1.714
97	3.214E-05	2.107E-05	3.667E-05	1.141	1.740
98	3.229E-05	2.083E-05	3.684E-05	1.141	1.768
99	3.244E-05	2.058E-05	3.700E-05	1.141	1.798
100	3.260E-05	2.033E-05	3.717E-05	1.140	1.829

Table 4. Diffusion properties at different temperatures based on algebraic formulae in **Table 1**.

Type/name	Mathematical form	Supplemental formula(e)	Reference
Rotational energy	$G\left(\frac{1}{\lambda_r}, T\right) = \frac{N_{J,\tau}}{N} = \frac{g_{J,\tau}\exp\left(-e_{J,\tau}/kT\right)}{Z_r}$	$g_{J,\tau} = \begin{cases} 3(2J+1), & if \quad \tau = odd \\ 2J+1, & if \quad \tau = even \end{cases}$ $Z_r = \sum g_{J,\tau}\exp(-e_{J,\tau}/kT)$	[7]
Translational energy	$F\left(\frac{1}{\lambda_t}, T\right) = \frac{dN}{Nd\left(1/\lambda_t\right)} = \frac{2}{\sqrt{\pi}}\frac{hc}{kT}\left(\frac{hc}{kT}\frac{1}{\lambda_t}\right)^{1/2}\exp\left(-\frac{hc}{kT}\frac{1}{\lambda_t}\right)$		[8]
Convolution, wave number basis	$H\left(\frac{1}{\lambda_{t,r}}, T\right) = \sum_{1/\lambda_r}^{\leq 1/\lambda_{t,r}} F\left[\left(\frac{1}{\lambda_{t,r}} - \frac{1}{\lambda_r}\right), T\right] * G\left(\frac{1}{\lambda_r}, T\right)$	$\int_0^{\infty} H_{1/\lambda_{t,r}} d(1/\lambda_{t,r}) = 1$	[1]
Convolution, wavelength basis	$H_{\lambda_{t,r}} = \frac{H_{1/\lambda_{t,r}}}{\lambda_{t,r}^2}$	$H_{1/\lambda_{t,r}} d(1/\lambda_{t,r}) = -H_{\lambda_{t,r}} d\lambda_{t,r}$	
Energy release	$\Delta e = \left(e_t - \frac{3}{2}kT\right) + \left(e_{J,\tau} - \frac{3}{2}kT\right) - H_{HB}$	H_{HB} = −23324.4 J/mol	[9]
Wave number basis	$\frac{1}{\lambda} = \frac{1}{\lambda_t} + \frac{1}{\lambda_r} + \frac{1}{\lambda_{cut}} = \frac{e_t}{hc} + \frac{e_{J,\tau}}{hc} + \frac{1}{hc}(-H_{HB} - 3kT)$	$\frac{1}{\lambda_{cut}} = \frac{1}{hc}(-H_{HB} - 3kT)$	[1]
Evaporation flux related to absorption	$\left(N_g''\right)_{abs} = \frac{1}{2}\int_0^{\infty} c_{12} Y_\lambda'' L_\lambda d\lambda$	$Y_{1/\lambda}'' = \sum_i \int_{\omega_i} \left(\frac{I_{\lambda,i}}{hc/\lambda}\cos\theta_i\right) d\omega_i$	
Evaporation flux related to spontaneous emission	$\left(N_g''\right)_{spon} = -\int_0^{\infty} a_{21} Z'' H_\lambda d\lambda$	$Z'' = \frac{1}{4} n_g \overline{u}$, $\overline{u} = \sqrt{\frac{8kT}{\pi m_g}}$	[8]
Evaporation flux related to induced emission	$\left(N_g''\right)_{ind} = -\int_0^{\infty} c_{21} X_\lambda'' H_\lambda d\lambda$	$X_\lambda'' = \frac{1}{2}\left(n_g \pi r_g^2\right)\sum_i \left[\alpha r_g \int_{\omega_i} \frac{I_{\lambda,i}}{hc/\lambda}(\cos\theta_i) d\omega_i\right]$, $\alpha \sim O$ [1]	
Einstein's relation	$a_{21}\left(\frac{\overline{n_g}\,\overline{u}}{4\pi I_{b,\lambda}(hc/\lambda)^{-1}}\right)H_\lambda = c_{12}L_\lambda - c_{21}\left(\alpha\overline{n_g}\pi r_g^3\right)H_\lambda$	$I_{b,\lambda} = \frac{2hc^2}{e^{(hc/\lambda kT)}-1}\left(\frac{1}{\lambda}\right)^5\left[\frac{W}{m^2\text{-}sr\text{-}\mu m}\right]$	
	$I_{b,1/\lambda} = \frac{2hc^2}{e^{(hc/\lambda kT)}-1}\left(\frac{1}{\lambda}\right)^3\left[\frac{W}{m^2\text{-}sr\text{-}cm^{-1}}\right]$	$I_{b,\nu} = \frac{2h\nu^3}{c^2\left(e^{h\nu/kT}-1\right)}\left[\frac{W}{m^2\text{-}sr\text{-}s^{-1}}\right]$	

http://dx.doi.org/10.5772/67401

Type/name	Mathematical form	Supplemental formula(e)	Reference
Blackbody radiation intensity			
Net evaporation flux	$\left(N_g''\right)_\lambda = \frac{1}{2}\left(\frac{I_\lambda d\lambda}{hc/\lambda}\cos\theta\delta\omega\right)\left(a_{21}\frac{\overline{n_g}\,\overline{u}}{4\pi I_{b,\lambda}(hc/\lambda)^{-1}}H_\lambda\right) - \frac{1}{2}s\left(\frac{I_{b,\lambda}d\lambda}{hc/\lambda}2\pi\right)\left(a_{21}\frac{\overline{n_g}\,\overline{u}}{4\pi I_{b,\lambda}(hc/\lambda)^{-1}}H_\lambda\right)$	$\left\lvert a_{21}\left(\frac{\overline{n_g}\,\overline{u}}{4\pi I_{b,\lambda}(hc/\lambda)^{-1}}\right)H_\lambda\right\rvert \gg \left\lvert c_{21}\left(\alpha\overline{n_g}\pi r_g{}^3\right)H_\lambda\right\rvert$	

Table 5. Population distribution functions for water-vapor molecules and equations related to phase-transition radiation.

2.4.2. *Characteristic wavelength*

Population distributions in **Figure 4** suggest the active spectral range for characteristic radiation. For $T = 0$–$100°C$ characteristic wavelength peaks at 5–6 μm in the mid-IR range, which is outside the visible spectrum. **Figure 4** is a duplicate of wavelength-based population distributions *H(λ, T)* in [1] for vapor molecules with respect to the transition with one hydrogen bond (*HB*) breaking/formation. The integration of the population distribution function over the entire wavelength range is unity by definition and is confirmed through numerical tests.

2.4.3. *Evaporation flux equations*

Phase-transition radiation involves three energy transition modes when photons interact with water molecules on the vapor-liquid interface: absorption, spontaneous emission, and induced emission. Evaporation fluxes related to these three modes for vapor molecules, N_g'' [#/m^2-s], can be associated with phase-transition radiation through (1) collision rates of water molecules and photons at the interface and (2) Einstein's coefficients. To avoid lengthy descriptions, evaporation flux equations are tabulated in **Table 5**. The spirit of these equations is that when photons interact with water molecules at the interface, there are chances for phase transitions

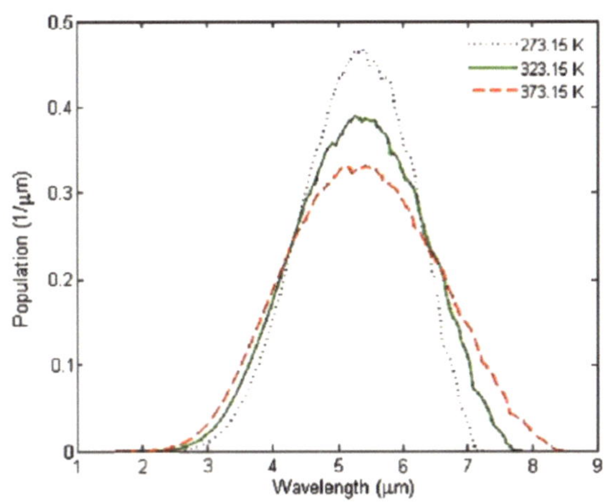

Figure 4. Population distribution functions of vapor molecules at different temperatures (duplicated from [1]).

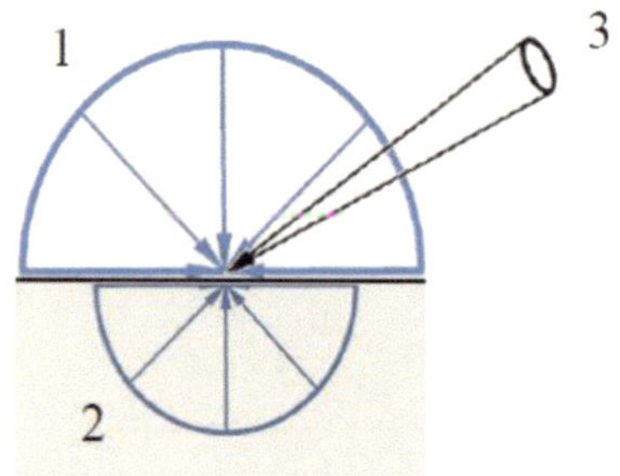

Figure 5. Different photon sources (1: external, diffuse; 2: internal, diffuse; 3: external, collimated) on the interface.

to happen. These possibilities are described as Einstein's coefficients for absorption (c_{12}), spontaneous emission (a_{21}), and induced emission (c_{21}).

Figure 5 shows three typical radiation sources on the vapor-liquid interface: (1) external diffuse-radiation $I_{\lambda,1}$; (2) internal diffuse-radiation $I_{\lambda,2}$; and (3) external collimated-radiation $I_{\lambda,3}$. Einstein's relation in **Table 5** is obtained by letting $I_{\lambda,1} = I_{\lambda,2} = I_{b,\lambda}$ (Blackbody radiation) and $I_{\lambda,3} = 0$.

2.4.4. Supersaturation by characteristic radiation

Assuming ideal gas for water vapor and using overbar to indicate saturation conditions, vapor number density n_g is related to supersaturation s via, $\boldsymbol{n_g = \overline{n_g}(1+s)}$. Saturated air with 100% relative humidity is equivalent to "zero" supersaturation.

2.4.5. Net evaporation flux

For semi-infinite water exposed to (1) Blackbody radiation $I_{b,\lambda}$ at the same temperature as water surface and (2) collimated radiation I_λ with solid angle $\delta\omega$ at angle θ (see **Figure 5** with $I_{\lambda,1} = I_{\lambda,2} = I_{b,\lambda}$ and $I_{\lambda,3} = I_\lambda$), invoking Einstein's relation, the net evaporation flux [#/m^2-s] for $\lambda \to \lambda + d\lambda$ is,

$$\begin{aligned}\left(N''_g\right)_\lambda = &\frac{1}{2}\left(\frac{I_\lambda d\lambda}{hc/\lambda}\cos\theta\delta\omega\right)\left\{\left(a_{21}\frac{\overline{n_g}\,\overline{u}}{4\pi I_{b,\lambda}(hc/\lambda)^{-1}}H_\lambda\right) - sc_{21}\left(\alpha\overline{n_g}\pi r_g{}^3\right)H_\lambda\right\} \\ &-\frac{1}{2}s\left(\frac{I_{b,\lambda}d\lambda}{hc/\lambda}2\pi\right)\left\{\left(a_{21}\frac{\overline{n_g}\,\overline{u}}{4\pi I_{b,\lambda}(hc/\lambda)^{-1}}H_\lambda\right) + c_{21}\left(\alpha\overline{n_g}\pi r_g{}^3\right)H_\lambda\right\}.\end{aligned} \tag{26}$$

The collimated radiation I_λ can be regarded as an excess radiation in addition to the background Blackbody radiation at thermodynamic equilibrium. If I_λ is taken away from Eq. (26), the net evaporation flux will automatically vanish to satisfy thermodynamic equilibrium conditions. Since s is a finite number, a simple form for evaporation flux is obtained assuming that spontaneous emission dominates the emission contribution in Einstein's relation (which is based on thermodynamic equilibrium conditions), giving,

http://dx.doi.org/10.5772/67401

$$\left(N''_g\right)_\lambda = \frac{1}{2}\left(\frac{I_\lambda d\lambda}{hc/\lambda}\cos\theta\delta\omega\right)\left(a_{21}\frac{\overline{n_g}\overline{u}}{4\pi I_{b,\lambda}(hc/\lambda)^{-1}}H_\lambda\right) - \frac{1}{2}s\left(\frac{I_{b,\lambda}d\lambda}{hc/\lambda}2\pi\right)\left(a_{21}\frac{\overline{n_g}\overline{u}}{4\pi I_{b,\lambda}(hc/\lambda)^{-1}}H_\lambda\right). \tag{27}$$

2.4.6. *Supersaturation = 0*

Local thermodynamic equilibrium (*LTE*) with $s = 0$ can be assumed when the collimated incident radiation I_λ is much weaker than Blackbody radiation $I_{b,\lambda}$. This leads to a much simpler form for the net evaporation flux in $\lambda \rightarrow \lambda + d\lambda$,

$$\left(N''_g\right)_\lambda = \frac{1}{2}\left(\frac{I_\lambda d\lambda}{hc/\lambda}\cos\theta\delta\omega\right)\left(a_{21}\frac{\overline{n_g}\overline{u}}{4\pi I_{b,\lambda}(hc/\lambda)^{-1}}H_\lambda\right). \tag{28}$$

The resulting evaporation flux is expected to be small. The surface absorption efficiency f_s can be defined as the ratio of the net evaporation flux to the incident photon flux of the collimated radiation [6],

$$f_s = a_{21}\frac{\overline{n_g}\overline{u}}{4\pi I_{b,\lambda}(hc/\lambda)^{-1}}H_\lambda\left(= \frac{a_{21}H_\lambda}{4\pi^2 I_{b,\lambda}(hc/\lambda)^{-1}}\overline{n_g}\sqrt{\frac{8\pi kT}{m_g}}\right). \tag{29}$$

Eq. (29) can also be interpreted as the fraction of radiation absorbed at interface [2].

Einstein's coefficient of spontaneous emission a_{21} is a small number on the order of 10^{-7} to 10^{-8} [10, 11]. With $a_{21} = 3 \times 10^{-8}$ [10] Eq. (29) is plotted in **Figure 6**. Readers shall not be bothered by the peaks of f_s between 2.5 and 3 μm because the evaporative flux in Eq. (28) is predominantly subject to the population distribution function H_λ.

2.4.7. *Supersaturation ≠ 0*

If the collimated radiation is of moderate strength (i.e., not large enough to break *LTE*), absorption of excess radiation may result in elevated evaporation flux and thus supersaturation. Eq. (27) in $\lambda \rightarrow \lambda + d\lambda$ can be written in terms of f_s,

$$\left(N''_g\right)_\lambda = \frac{1}{2}f_s\left(\frac{I_\lambda d\lambda}{hc/\lambda}\cos\theta\delta\omega\right) - \frac{1}{2}sf_s\left(\frac{I_{b,\lambda}d\lambda}{hc/\lambda}2\pi\right). \tag{30}$$

2.4.8. *Quasi-steady supersaturation*

Relation between supersaturation and incident radiation can be established for a thermodynamic system at quasi-steady state. Consider such a system originally at quasi-steady state without excess IR radiation to drive out water molecules from the surface and that there is neither curvature effects nor salutes to exert additional influences on equilibrium pressure. Vapor pressure at interface is just the saturation pressure at the surface temperature. As soon as an additional mid-IR radiation field is applied to the system, the evaporation rate begins to

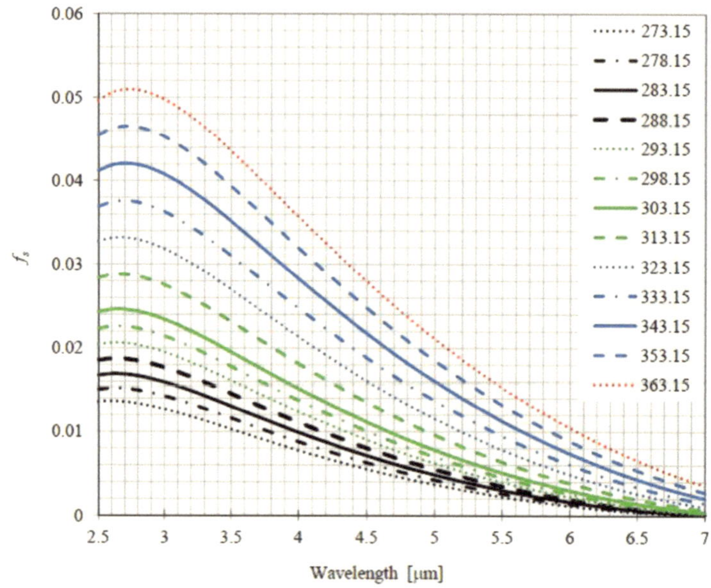

Figure 6. Surface absorption efficiency f_s.

exceed the condensation rate, and more vapor pressure starts to build up on water surface before vapor diffuses away. This is the onset of supersaturation.

As a result of the elevated vapor pressure on the surface, molecular diffusion is boosted by the increased vapor concentration gradient, leading to a higher evaporation rate. Consequently, surface temperature drops in response to the enhanced evaporation, which takes heat away from all possible sources (such as air, water, and radiation sources in all available spectral ranges) as the system tries to move toward another equilibrium state. Since saturation pressure is strongly dependent on temperature, the degree of supersaturation will be further lifted up as the surface temperature drops. Eventually supersaturation will increase to reach a new quasi-steady state at which absorption of characteristic radiation ("+" term) can no longer surpass emission ("−" term) in Eq. (30). The new quasi-steady supersaturation is assumed to be,

$$s = \frac{\cos\theta\delta\omega}{2\pi}\int_0^\infty \frac{I_\lambda}{I_{b,\lambda}} H_\lambda d\lambda, \tag{31}$$

which gives a zero evaporation flux in Eq. (30) for the integration value over the entire spectral range. Note that I_λ here is the excess radiation intensity and the quasi-steady s is independent of the probability constant a_{21}. For diffuse radiation with intensity I_λ coming from above (i.e., hemispherical radiation intensity independent of incident angle), the quasi-steady supersaturation is,

$$s = \frac{1}{2}\int_0^\infty \frac{I_\lambda}{I_{b,\lambda}} H_\lambda d\lambda. \tag{32}$$

Supersaturations caused by background Blackbody radiation with temperature T_{bkgd} is plotted in **Figure 9** following Example 5. The elevated supersaturation due to excess IR radiation enhances evaporation by uplifting the surface vapor mass fraction, $m_{1,s}$, in Eq. (19). In distillation applications, this mechanism allows enhanced evaporation to take place below the boiling point.

In Eqs. (26)–(32), (H/I_b) is dependent on temperature but invariant with respect to the evaluation basis (wave number $1/\lambda$ or frequency ν, or even wavelength λ). To facilitate engineering analysis, coefficients for curve-fits of (H/I_b) defined in Eq. (33) are tabulated in **Table 6** for λ from 2.5 μm to the far wing where H_λ approaches zero (see **Figure 4**, generally >7.25 μm, depending on T). The units for λ and (H/I_b) in Eq. (33) are, respectively, [μm] and [cm²-sr/W].

$$\frac{H}{I_b}\left[\frac{cm^2\text{-}sr}{W}\right] = \sum_{i=0}^{6} P(i)\left(\lambda\left[\mu m\right]\right)^i. \tag{33}$$

2.4.9. *Local stability of quasi-steady supersaturation*

The quasi-steady supersaturation in Eq. (31) or (32) is in favor of the local stability of the thermodynamic system. This is explained as follows. For a system already at quasi-steady state under IR characteristic radiation, a perturbation is added to the established supersaturation to examine the stability of the system. The perturbation can be either positive or negative with respect to the quasi-steady supersaturation. For a positively perturbed supersaturation, the net radiation-induced evaporation rate in Eq. (30) will become negative as a result of the slightly increased supersaturation to bring the system back to its original quasi-steady state.

T (°C)	P(6)	P(5)	P(4)	P(3)	P(2)	P(1)	P(0)
0	−1.8049E+00	7.0424E+01	−1.1017E+03	9.0590E+03	−4.1597E+04	9.8249E+04	−8.2804E+04
5	−2.3238E+00	8.2720E+01	−1.2058E+03	9.3386E+03	−4.0726E+04	9.2567E+04	−7.7111E+04
10	−1.6147E+00	5.8630E+01	−8.6986E+02	6.8497E+03	−3.0353E+04	6.9958E+04	−5.8710E+04
15	−1.1467E+00	4.2407E+01	−6.3940E+02	5.1118E+03	−2.2981E+04	5.3643E+04	−4.5323E+04
20	−8.2035E-01	3.0924E+01	−4.7402E+02	3.8482E+03	−1.7552E+04	4.1491E+04	−3.5292E+04
25	−5.9907E-01	2.2972E+01	−3.5735E+02	2.9410E+03	−1.3589E+04	3.2489E+04	−2.7806E+04
30	−4.4173E-01	1.7229E+01	−2.7193E+02	2.2681E+03	−1.0613E+04	2.5656E+04	−2.2090E+04
40	−2.4969E-01	1.0042E+01	−1.6270E+02	1.3902E+03	−6.6559E+03	1.6420E+04	−1.4297E+04
50	−1.1513E-01	5.0540E+00	−8.7390E+01	7.8839E+02	−3.9565E+03	1.0133E+04	−8.9693E+03
60	−8.4490E-02	3.6328E+00	−6.2201E+01	5.5868E+02	−2.8027E+03	7.2032E+03	−6.4134E+03
70	−4.8756E-02	2.1939E+00	−3.8903E+01	3.6015E+02	−1.8567E+03	4.8836E+03	−4.4002E+03
80	−2.8232E-02	1.3386E+00	−2.4662E+01	2.3577E+02	−1.2505E+03	3.3683E+03	−3.0713E+03
90	−1.6512E-02	8.2929E-01	−1.5903E+01	1.5708E+02	−8.5720E+02	2.3644E+03	−2.1813E+03

Table 6. Curve-fit coefficients for (H/I_b) in Eq. (33).

On the contrary, a negative perturbation for supersaturation results in a positive net radiation-induced evaporation rate to restore the system to its prior unperturbed quasi-steady state. Therefore a locally stable state is established at the quasi-steady supersaturation.

2.5. Materials and methods

The most crucial part in the engineering application of radiation-enhanced evaporation is the selection of the IR radiation source. Since the characteristic radiation is in the mid-IR spectral range, lasers and LEDs designed for this range fit in well for this purpose. However, they are not economically feasible on industrial scales. An alternative way of generating mid-IR radiation is to heat up a Blackbody-like material to a desired temperature such that the peak of its Blackbody-like radiation locates within the spectral range of characteristic radiation. Based on Wien's displacement law (λ_{max} [μm] × T [K] = 2898 [μm-K]), T = 200–300°C corresponds to λ_{max} = 6.12–5.06 μm. At T = 250°C, the corresponding λ_{max} (=5.54 μm) appears to be a preferred choice because it is near the maximums of population distributions in **Figure 4**.

Surface treatment is perhaps one of the most affordable methods to create Blackbody-like surfaces. Applying paints over surfaces can achieve this purpose because paints are usually "black" in the mid-IR spectral range. Anodized aluminum coating, which is commonly practiced in industries to passivate aluminum surfaces, is able to produce a moderately high emittance (~0.85, depending on several factors such as temperature, color, thickness, and roughness) [12] to somewhat resemble Blackbody surfaces. If there is no appearance preference in the engineering design, black color is generally recommended for surface treatment to mimic the Blackbody emittance.

As far as mass and heat transfer analysis is concerned, the temperature at water surface T_s needs to be determined before other thermal variables can be calculated. This can be done by plotting Eq. (25) in spreadsheet to visually determine T_s, as shown in Example 3. Alternatively, a computational method for computer-aided calculations is suggested in **Figure 7**. The logic behind this method for finding T_s is explained also in the same example.

2.6. Results and discussion

This section guides readers through examples to deal with water vaporization problems in distillation applications. Comments are made following the results of these examples to help readers understand methodologies presented in this chapter.

2.6.1. Example 1—isothermal water under mid-IR radiation

Consider a still lake with surface temperature T_s fixed at 20°C and a temperature gradient that vanishes at the bottom of the lake. Provided that a diffuse mid-IR radiation field with $q_{r,s}$ = 1000 W/m^2 is applied to water surface from above, estimate the maximum temperature change from its surface to the bottom due to radiative heating. Assume cloudy sky to skip solar radiation and evaporation flux, $\dot{m}'' = 10^{-4}$kg/m^2-s.

http://dx.doi.org/10.5772/67401

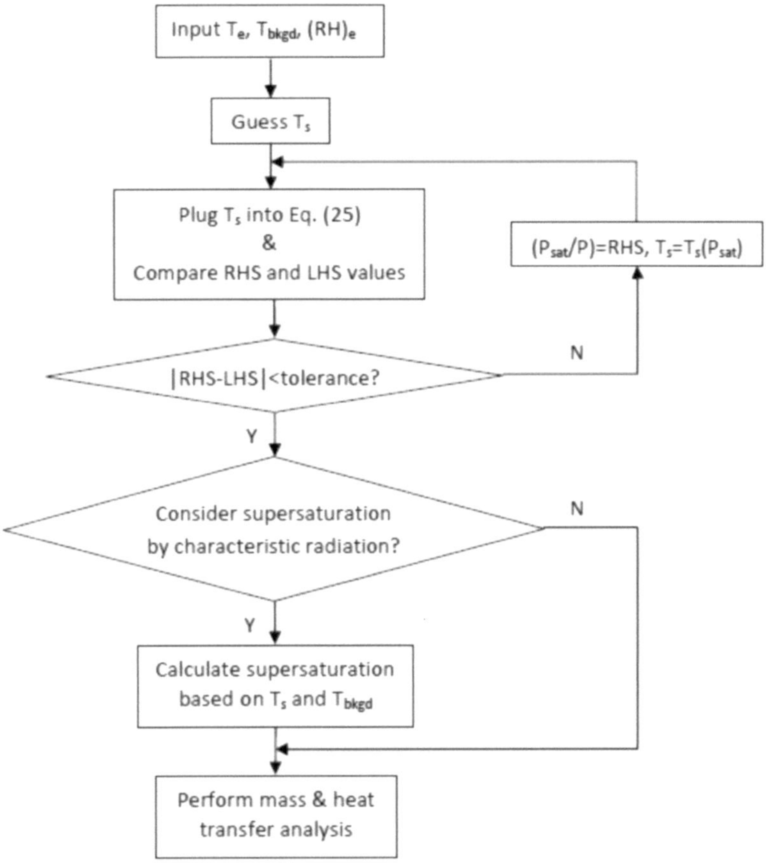

Figure 7. A flowchart for computer-aided calculations.

Solution

T_s= 20°C: k_l = 0.6 W/m-K, C= 4.2 kJ/kg-K, $\dot{m}''$= 10^{-4} kg/m^2-s => $\delta_t = k_l / (\dot{m}''C)$ = 1.42 m >>δ_r ~ 20 μm.

$0 \leq f_s \leq 1$, $q_{r,s}$ = 1000 W/m^2 => $\Delta T_{\max} \approx (1 - f_s)q_{r,s}\delta_r/k_l \leq 0.033°C$.

$$\therefore max(T - T_s) \approx \Delta T_{\max}\left(-e^{-K_a x} + 1\right)\Big|_{x\to\infty} = \Delta T_{\max} \leq 0.033°\mathrm{C}.$$

The maximum temperature change from water surface to the bottom is estimated to be 0.033°C increase for $q_{r,s}$ = 1000 W/m^2.

This example shows that, without other heating sources, water exposed to moderate strength of mid-IR radiation is essentially isothermal at quasi-steady state. A direct conclusion of the simplified equation is that water can be treated isothermal ($T_x = T_s$), regardless of the value of surface absorption coefficient f_s.

For distillation applications, IR radiation sources can be designed to primarily cover the spectral range of characteristic radiation. The external IR radiation fields are intended to be used as means to generate supersaturations instead of as major heat sources. The idea here is to apply the right amount of IR radiation in the right spectral range so that the desired supersaturation can be achieved in an economical way.

2.6.2. Example 2—economic evaluation

Given that the cost of household electricity is \$0.2 USD/kWh (for example in Cambridge, Massachusetts, USA) and the desired evaporation flux is 10^{-4} kg/m^2-s, evaluate the minimum cost (\$USD/Ga) for water distillation at 100°C. In the economic evaluation ignore heat loss/recovery, sensible heat for water to reach 100°C and costs related to vapor condensation and water collection.

Solution

Heat of vaporization at 100°C: 2257 kJ/kg.

Water density at 100°C: 958 kg/m^3.

Gallon to cubit meter conversion: 0.003785 m^3/Ga.

Cost per kg: $\frac{\$0.2\text{USD}}{\text{kWh}} \times \frac{2257\text{kJ}}{\text{kg}} \times \frac{\text{hour}}{3600\text{s}} = 0.125\text{USD/kg}.$

Cost per gallon: $\frac{\$0.125\text{USD}}{\text{kg}} \times \frac{958\text{kg}}{\text{m}^3} \times \frac{0.003785\text{m}^3}{\text{Ga}} = 0.453\ \text{USD/Ga}.$

Note that at room temperatures water density is 4% larger than that at 100°C, giving an adjusted minimum distillation cost \$0.47 USD/Ga. The retail price of distilled water at Walmart (brand: Great Value Distilled Water, 1 Gal) is \$0.88 USD/Ga. There are several options to reduce the production cost of distilled water: choosing a cheaper energy source, minimizing heat loss while vaporizing water, recovering heat release during vapor condensation, *or* using a new method for water vaporization such as what is proposed in this chapter.

2.6.3. Example 3—computational methods

Given that the temperature at the top of vapor layer thickness is $T_e = 30°\text{C}$ and the corresponding relative humidity $(RH)_e = 0.8$ and ignoring radiation effects, calculate the quasi-steady surface temperature T_s at $P = 1$ atm. For vapor layer thickness $L = 1$ mm, what is the quasi-steady evaporation flux?

Solution

2.6.3.1. Surface temperature

Since radiative heating is ignored, the isothermal approximation is applicable for liquid-water, i.e., $h_{1,o} = h_{1,u} = 4.2T_s[°\text{C}]$, and there is no IR induced supersaturation, i.e., $(RH)_s = 1$.

In **Figure 2** Le for saturated air ($Le \approx 1.19$) can be used because of the high relative humidity. From steam tables, $P_{sat,e}$ (T_e = 30°C) = 0.0419 atm. To begin with we shall assume a dilute system, in which $C_p \approx 1$ kJ/kg-K. The value of $m_{1,e}$ supports this assumption,

http://dx.doi.org/10.5772/67401

$$m_{1,e} \approx \frac{(0.8)(0.0419)}{1.61 - 0.61(0.8)(0.0419)} = 0.0211.$$

Eq. (25) gives,

$$\frac{P_{sat,s}}{P} \approx (0.8 \times 0.0419) + \frac{1.61}{1.19}\left[\frac{30 - T_s(°C)}{2501 - 3.2T_s(°C)}\right].$$

Two methods are used to obtain T_s.

Method 1: The correct value of T_s can be found by plotting the right hand side (*RHS*) and left hand side (*LHS*) values for $P_{sat,s}/P$ (based on either steam tables or **Table 1** for P_{sat}) with respect to guessed T_s in spreadsheet. The results are shown in **Figure 8**. The *LHS* curve intersects with the *RHS* to give the correct $T_s = 27°C$.

Method 2: As shown in **Figure 7** the value of T_s needs to be guessed to find the corresponding $P_{sat,s}$. Plugging in the guessed T_s the *RHS* result is compared with the *LHS* for $P_{sat,s}/P$. If further guesses are needed, the *RHS* value of the present guess can be conveniently used as the *LHS* value of the new guess. This guessing method is applicable because the *RHS* and *LHS* curves come across each other at the right T_s as shown in **Figure 8**.

Saturation pressure P_{sat} and temperature T_s can be calculated using formulas for P_{sat} and T_{sat} in **Table 1**.

1st guess: $T_{sat,s} = 30°C \Rightarrow P_{sat,s}/P = 0.04189 \Rightarrow RHS = 0.03353 < 0.04189 = LHS$.

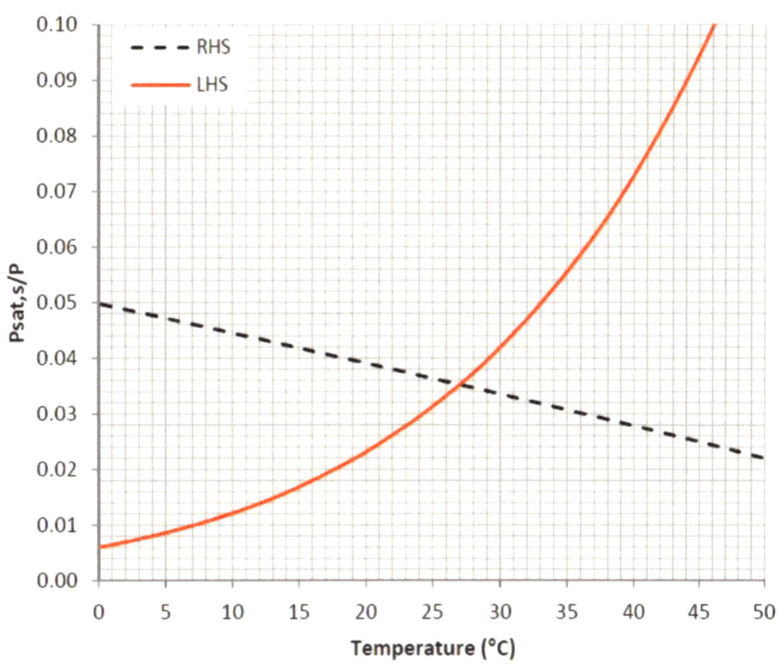

Figure 8. *RHS* and *LHS* curves for $P_{sat,s}/P$ in Example 3 with respect to guessed temperatures for T_s.

2nd guess: $P_{sat,s}/P = 0.03353 \Rightarrow T_{sat,s} = 26.18°C \Rightarrow RHS = 0.03566 < 0.03353 = LHS$.

3rd guess: $P_{sat,s}/P = 0.03566 \Rightarrow T_{sat,s} = 27.22°C \Rightarrow RHS = 0.03510 < 0.03566 = LHS$.

4th guess: $P_{sat,s}/P = 0.03510 \Rightarrow T_{sat,s} = 26.95°C \Rightarrow RHS = 0.03524 \approx 0.03510 = LHS$.

The 4th guess gives a fairly good match between *RHS* and *LHS*. For $T_s = 27°C$, the mass fraction of vapor at surface $m_{1,s}$ is 0.0222, which also supports the dilute assumption. This method seems to be tedious but becomes powerful in computer-aided computations, in which the iteration method can easily be implemented in the code.

2.6.3.2. Evaporation flux

Table 1: for $T = T_{avg} = (T_e + T_s)/2 = 28.5°C \Rightarrow \rho = 1.17\ \text{kg/m}^3$, $D = 2.60\times10^{-5}\ \text{m}^2/\text{s}$.

$$\dot{m}'' \approx \rho D \frac{m_{1,s} - m_{1,e}}{L} \approx (1.17\text{kg/m}^3)(2.60 \times 10^{-5}\text{m}^2/\text{s})\left(\frac{0.0222 - 0.0211}{0.001m}\right) = 3.35 \times 10^{-5}\text{kg/m}^2\text{-s}$$

The calculated T_s (=27°C) is quite close to the wet-bulb temperature for the high *RH* condition in this example. Based on the online *RH* calculator by National Weather Service Weather Forecast Office (http://www.srh.noaa.gov/epz/?n=wxcalc_rh), the web-based wet-bulb temperature is 27.13°C.

2.6.4. Example 4—enhanced evaporation by supersaturation

Following Example 3, assume that surface supersaturation can be elevated to 20% by mid-IR radiation for the isothermal liquid, and calculate the quasi-steady surface temperature and evaporation flux. Assume radiative heating in the mid-IR range can be neglected, i.e., $T_{r,s} << (T_e - T_s)$, in this example (for practical situations see Example 5).

Solution

2.6.4.1. Surface temperature

To approach the right value of T_s Eq. (25) is employed with $(RH)_s = 1.2$ and $T_{r,s} = 0$. **Table 1** is used to calculate P_{sat} and T_{sat} at surface. *Le* is approximated by 1.19 for this high *RH* system. Based on Method 2 in Example 3 (also shown in **Figure 7**), surface temperature can be evaluated to give $T_s = 24.56°C$. The quasi-steady surface temperature is brought down by ~2.5°C as compared to the saturated water surface in Example 3.

2.6.4.2. Evaporation flux

Table 3: for $T_s = 24.56°C \Rightarrow m_{1,e} = 0.0211$, $m_{1,s} = 0.0230$.

Table 1: for $T = T_{avg} = (T_e + T_s)/2 = 27.28°C \Rightarrow \rho = 1.17\ \text{kg/m}^3$, $D = 2.58\times10^{-5}\ \text{m}^2/\text{s}$.

$$\dot{m}'' \approx \rho D \frac{m_{1,s} - m_{1,e}}{L} \approx (1.17\text{kg/m}^3)(2.58 \times 10^{-5}\text{m}^2/\text{s})\left(\frac{0.0230 - 0.0211}{0.001m}\right) = 5.74 \times 10^{-5}\text{kg/m}^2\text{-s}$$

By boosting supersaturation to 20%, evaporation mass flux is 70% higher than that (=3.35×10^{-5} kg/m^2-s) in Example 3. The amount of mass flux may not seem appealing in this high *RH* example. Nevertheless, through this pure diffusion problem it is shown that supersaturation significantly enhances evaporation.

2.6.5. *Example 5—engineering application of induced supersaturation for water vaporization*

Consider a still indoor pool with uniform water temperature 293.15K and the temperature of the ceiling is the same as the ambient at 303.15K. Assume Blackbody for the ceiling and water surface, and calculate the quasi-steady supersaturation and the relative humidity above vapor layer. For vapor layer thickness $L = 1$ mm, compute the evaporation flux and analyze how heat is supplied to liquid-water during the evaporation process. If radiation energy from the ceiling is supplied by electric heating, what is the unit cost for water vaporization based on the electricity price in Example 2 (i.e., $0.2 USD/kWh)?

Solution

2.6.5.1. *Supersaturation*

The quasi-steady supersaturation is evaluated using Eq. (32) with the background Blackbody temperature $T_{bkgd} = 303.15$K and the surface Blackbody temperature $T_s = 293.15$K,.

$$s = \frac{1}{2}\int_0^{\infty} \frac{I_{b,\lambda}(T_{bkgd}) - I_{b,\lambda}(T_s)}{I_{b,\lambda}(T_s)} H_\lambda(T_s) d\lambda. \tag{34}$$

Using **Table 5** for H_λ, the integration result for s is $s = 0.185$. A slightly different value of $s = 0.187$ is obtained based on H/I_b curve-fit values from **Table 6**.

In the midst of various sources of experimental uncertainties, the elevated supersaturation can easily be misidentified as a different thermodynamic variable such as a higher surface temperature. It is shown in the next example that, without considering Eq. (32) or (34), increasing T_s by 2°C gives $s = 0$ to satisfy mass and heat transfer equations at a different evaporation rate. This temperature difference (2°C) is within the accuracy range of K-type thermocouples [13].

2.6.5.2. *Relative humidity*

The radiation characteristic temperature $T_{r,s}$ is taken into account to solve for the value of $(RH)_e$.

Table 1: for $T_{avg} = 25°\text{C} \Rightarrow k = 2.59 \times 10^{-2}$ W/m-K (curve-fit coefficients in **Table 2** [5]), and

$$T_{r,s} = \frac{q_{r,s}L}{k} = \frac{\sigma(T_e^4 - T_s^4)L}{k} = \frac{60 \times 0.001}{0.0259} = 2.32 \ [\text{K}]$$

Using Method 1 from Example 3 with $(RH)_s = 1 + s = 1.19$ and $Le = 1.19$, the relative humidity above vapor layer is $(RH)_e = 0.49$.

2.6.5.3. *Mass and heat transfer*

Vapor mass fractions at the s- and e-states are, respectively, $m_{1,s} = 0.01724$ and $m_{1,e} = 0.01286$. The dilute approximation is valid. For $T_{avg} = (303.15\text{K} + 293.15\text{K})/2 = 25°\text{C}$, the corresponding evaporation flux is

$$\dot{m}'' \approx \rho D \frac{m_{1,s} - m_{1,e}}{L} \approx (1.18)\left(2.55 \times 10^{-5}\right)\left(\frac{0.01724 - 0.01286}{0.001}\right) = 1.318 \times 10^{-4}\left[\mathrm{kg/m^2\text{-}s}\right].$$

From steam tables, heat of evaporation: h_{fg} (20°C) = 2453 [kJ/kg].

Total heat flux required for evaporation: $H_{fg} = \dot{m}'' \times h_{fg} = 323$ [W/m^2].

Since the liquid is essentially isothermal under mid-IR radiation, Eqs. (8) and (21) can be combined to give,

$$\dot{m}'' h_{fg} = -q_{c,u} + f_s q_{r,s} + k\left.\frac{dT}{dy}\right|_s = \dot{m}''(h_{1,o} - h_{1,u}) + q_{r,s} + k\left.\frac{dT}{dy}\right|_s = q_{r,s} + k\left.\frac{dT}{dy}\right|_s.$$

The conduction contribution to H_{fg} is

$$k\left.\frac{dT}{dy}\right|_s = 0.0259 \times \frac{30 - 20}{0.001} = 259\left[W/m^2\right].$$

The radiation term $q_{r,s}$ from (b) is 60 W/m^2. Most of energy required for evaporation, 80% (=259/323), is conducted from the air.

2.6.5.4. Economic evaluation

Radiation energy per unit mass: $\frac{60\mathrm{W}}{\mathrm{m}^2} \div \frac{1.318\times10^{-4}\mathrm{kg}}{\mathrm{m}^2\text{-s}} = 455.2\ \mathrm{kJ/kg}$

Water density at 20°C: 998 kg/m^3.

Gallon to cubit meter conversion: 0.003785 m^3/Ga.

Cost per kg: $\frac{\$0.2\mathrm{USD}}{\mathrm{kWh}} \times \frac{455.2\mathrm{kJ}}{\mathrm{kg}} \times \frac{\mathrm{hour}}{3600\mathrm{s}} = 0.0253\ \mathrm{USD/kg}$

Cost per gallon: $\frac{\$0.0253\mathrm{USD}}{\mathrm{kg}} \times \frac{998\mathrm{kg}}{\mathrm{m}^3} \times \frac{0.003785\mathrm{m}^3}{\mathrm{Ga}} = 0.096\ \mathrm{USD/Ga}$.

As compared to Example 2, the unit cost (per gallon) for water vaporization is reduced from \$47 cents to \$10 cents. In practical engineering applications, the ceiling temperature, which reflects T_{bkgd} for the radiation source, can be different from the ambient temperature and the production volume can be scaled up by increasing water surface area.

Eq. (34) describes general situations for quasi-steady supersaturations induced by background (hemispherical) Blackbody radiation sources. **Figure 9** is constructed based on Eq. (34) to give supersaturations for some representative surface temperatures T_s and background Blackbody temperatures T_{bkgd}. It is seen in **Figure 9** that a higher T_s corresponds to a lower quasi-steady s for a fixed T_{bkgd}, which is the upper limit of T_s to give s = 0. In order to achieve s = 0.2, T_{bkgd} needs to be higher than T_s by 10.7–13.0°C respectively for T_s = 20–50°C. For a fixed T_s at 20°C, s can be uplifted from 0.2 to 0.6 by increasing T_{bkgd} from 30.7 to 46.0°C.

Note that a recommended Blackbody temperature to induce supersaturation is 250°C as discussed in Section 2.5. Since it is costly to attain such a high temperature for the entire hemispherical surface above water surface on industrial scales, a radiation source with this

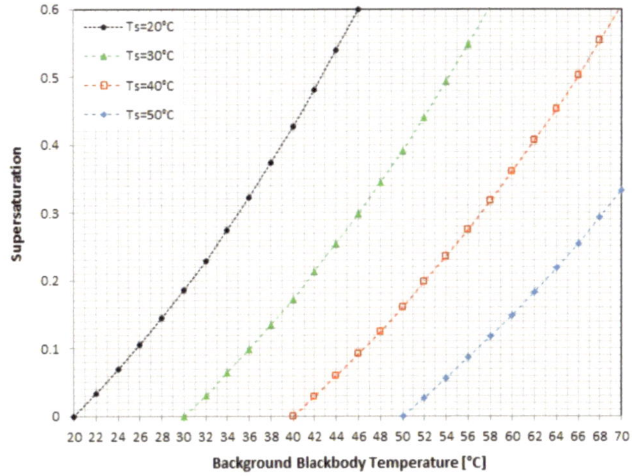

Figure 9. Quasi-steady supersaturations caused by diffuse background radiation sources characterized by Blackbody temperatures T_{bkgd} for surface temperatures T_s = 20–50°C.

temperature can be combined with other radiation sources in the engineering design. Eq. (31) can be used to integrate contributions from all radiation sources.

2.6.6. Example 6—conventional engineering analysis ignoring phase-transition radiation

Following Example 5, with the same ambient temperature, ceiling temperature, and $(RH)_e$ (=0.49), assume saturated water for the pool surface, i.e., $s = 0$ or $(RH)_s = 1$, calculate the corresponding surface temperature, evaporation flux, and the amount of heat conducted from air to the pool surface. In this example the effect of characteristic radiation on supersaturation is ignored.

Solution

2.6.6.1. Surface temperature

Surface temperature T_s is expected to be slightly higher than that in Example 5 for a lower $(RH)_s$. From **Table 1**, for a slightly higher T_{avg}, a slightly higher $k = 2.6 \times 10^{-2}$ W/m-K (curve-fit coefficients in **Table 2** [5]) can be used. Using Method 1 from Example 3 with $(RH)_s = 1$, $(RH)_e$ = 0.49, and $Le = 1.19$ => $T_{r,s} = 1.87$°C and $T_s = 22$°C.

2.6.6.2. Mass and heat transfer

Vapor mass fractions at the s- and e-states are, respectively, $m_{1,s} = 0.01638$ and $m_{1,e} = 0.01286$. For T_{avg} = (30°C + 22°C)/2 = 26°C, the corresponding evaporation rate is

$$\dot{m}'' \approx \rho D \frac{m_{1,s} - m_{1,e}}{L} \approx (1.18)(2.56 \times 10^{-5}) \left(\frac{0.01638 - 0.01286}{0.001} \right) = 1.063 \times 10^{-4} \left[\text{kg/m}^2\text{-s} \right].$$

From steam tables, heat of evaporation: h_{fg} (22°C) = 2449 [kJ/kg].

Total heat flux needed for evaporation: $H_{fg} = \dot{m}'' \times h_{fg} = 260$ [W/m^2].

The conduction contribution to H_{fg} is

$$k\left.\frac{dT}{dy}\right|_s = 0.0260 \times \frac{30-22}{0.001} = 208\ \left[\mathrm{W/m^2}\right].$$

The radiation contribution is 49 W/m^2. The evaporation flux without supersaturation drops by ~20% as compared to that in Example 5 and the amount of heat conducted/extracted from air also drops by ~20%.

3. Conclusion

For semi-infinite water under mid-IR radiation, the liquid phase is essentially isothermal as a result of the pronounced IR absorption of radiation in liquid-water. When vapor mass fraction is small compared to the air, the dilute approximation can be made and equations for mass and heat transfer can be significantly simplified. Together with these simplified forms for dilute approximations, empirical forms for thermodynamic properties are also presented in this chapter to facilitate engineering analysis.

The situation of supersaturation induced by IR characteristic radiation suggests a potential way to enhance evaporation below the boiling point. Characteristic radiation applied on the water surface directly drives water molecules out from the surface to cause evaporation. When excess IR radiation is continually applied over the surface, the quasi-steady supersaturation can be reached to enhance evaporation. Since water is nearly isothermal under mid-IR radiation with moderate radiation strength, a significant amount of heat is extracted from the air to supply the latent heat of evaporation. Blackbody-like materials heated to ~250°C can serve as mid-IR radiation sources. Economical Blackbody-like materials include black anodized aluminum surfaces and metal surfaces painted in black. With supersaturation induced by characteristic radiation in the mid-IR range, this chapter offers an economical way to enhance evaporation in distillation applications.

Acknowledgements

The second author would like to acknowledge the support of the Hermia G. Soo Professorship and NSF Grant 1457128.

Author details

Kuo-Ting Wang[1,2]*, M. Quinn Brewster[1] and Wei-Hsiang Lai[2]

*Address all correspondence to: calebuiuc@gmail.com

1 Department of Mechanical Science and Engineering, University of Illinois at Urbana-Champaign, United States

2 Department of Aeronautics and Astronautics, National Cheng Kung University, Taiwan

http://dx.doi.org/10.5772/67401

References

[1] Wang K.T. and Brewster M.Q. Phase-transition radiation in vapor condensation process. Int. Commun. Heat Mass Transfer. 2010;37:945–949.

[2] Brewster M.Q. Evaporation and condensation of water mist/cloud droplets with thermal radiation. Int. J. Heat Mass Transfer. 2015;88:695–712.

[3] NASA. Earth fact sheet [Internet]. Available from: http://nssdc.gsfc.nasa.gov/planetary/factsheet/earthfact.html [Accessed: Aug 2016]

[4] NOAA. Vapor pressure [Internet]. Available from: http://www.srh.noaa.gov/images/epz/wxcalc/vaporPressure.pdf [Accessed: Aug 2016]

[5] Tsilingiris P.T. Thermophysical and transport properties of humid air at temperature range between 0 and 100C. Energy Conversion and Management. 2008;49:1098–1110.

[6] Wang K.T. On Phase-Transition Radiation of Water. 1st ed. Germany: Lambert Academic Publishing; 2012. 177 p. ISBN: 978-3-659-20514-9.

[7] Randall H.M., Dennison D.M., Ginsburg N. and Weber L.R. The far infrared spectrum of water vapor. Physical Review. 1937;52:160–174.

[8] Vincenti W.G. and Kruger C.H. Jr. Introduction to Physical Gas Dynamics. New York: John Wiley and Sons, Inc.; 1965.

[9] Suresh S.J. and Naik V.M. Hydrogen bond thermodynamic properties of water from dielectric constant data. J. Chem. Phy. 2000;113(21):9727–9732.

[10] Brewster M.Q., Wang K.T., Wu W.H. and Khan M.G. Temperature effect on phase-transition radiation of water. J. Heat Transfer. 2014;136(6):062704.

[11] Wang K.T. and Brewster M.Q. Phase-change radiation: Enhanced infrared absorption in vapor-liquid water mixtures. J. Thermophys. Heat Transfer. 2013;27(3):447–457.

[12] Kauder L. Spacecraft Thermal Control Coatings References. NASA/TP. 2005; 212792.

[13] Omega. Thermocouple Reference Data [Internet]. Available from: http://www.omega.com/prodinfo/thermocouples.html [Accessed: Aug 2016]

Chapter 6

Distillation Techniques in the Fruit Spirits Production

Nermina Spaho

Additional information is available at the end of the chapter

http://dx.doi.org/10.5772/66774

Abstract

During the distillation of the fermented fruit mash or juice, ethanol and water are the carriers of a huge number of the other volatile aroma compounds. Unique and distinctive flavour of the final spirits depends on their quantity and quality. Fruit spirits have higher concentration of almost all types of volatile compounds with comparing to other types of distilled spirits. The art of distillation run is to obtain the best balance between congeners present. Two different types of distillation equipment are used for the production of fruit spirits: copper Charentais alembic and batch distillation columns. Although both distillation methods are based on the same theoretical principles, a different quantity of the flavour compounds of the final spirits is produced by using different distillation equipment. The main difference was shown in different distributions of the methanol, n-propanol, higher alcohols and fatty acid esters. Distillation methods need to be adjusted for each fruit spirits regardless to distillation equipment employed because fermented mash of different fruit varieties has a different requirement for distilling. Alembic stills yield better aroma and more characteristic fruit distillates but are slow and require more labour. Column still cleans the distillate giving a decent aroma and higher concentration of alcohol.

Keywords: alembic, distillation column, fruit spirits, distillation cut, aroma compounds

1. Introduction

Fruit spirits are popular alcohol beverages due to their unique flavour. They are consumed in various countries all over the world. Very often fruit spirits represent the national drink of the country, so the Hungarians are proud on their apricot spirits; French on Calvados; Italians on apple spirits [1]; Greeks on mulberry spirits [2]; Balkan countries [3, 4] and Eastern European countries on plum spirits [5]; and Germans and Swiss on their 'kirschwasser' [6]. Although fruit spirits could be made from each type of fruit that contains at least a few amounts of sugar, either cultured or wild fruit, the most popular fruit spirits are made from plums, cherries, apples, pears, apricots and quinces. The nice, unique and pleasant aroma is common for all of them. The main ingredients of these beverages are water and ethanol, and they account around 99% of the total content of the spirits. Nevertheless, the fruit spirits are not a pure diluted ethanol; in that case one could not make differences between plum spirits from quince or pear spirits. Fruit spirits must be recognised on the raw material from which it obtained. Hundreds of different compounds, so-called congeners, have been identified in fruit spirits, present in very low concentration but crucial for the quality of beverages. The composition and concentrations of these congeners vary depending on the cultivar used for the production, fermentation procedure, yeast strain used, time of storage of fermented mash before distillations, distillation technique, use of post distillation processes and maturation of spirits. Thus, it seems that the quality of spirit and its identity and character are influenced by a huge number of factors. Some authors favoured raw material as a factor of great influence on character and authenticity of fruit spirits [7–9]; others said that fermentation is the most important activities in aroma compound production; another, however, wrote that distillation is a technique that manages the composition of aroma compounds present in the spirits [10–12], whilst some other authors highlighted the ageing of spirits in creating their unique quality [13, 14]. However, the truth is somewhere between of all these views. Nevertheless all researchers are agreed that all phases have to be adequate carried out in order to achieve the fruit spirits premium quality.

The production of fruit spirits has a long tradition in Croatia and the other West Balkan countries (Bosnia and Herzegovina, Serbia, Montenegro, Macedonia, Kosovo and Albania). The most popular is plum brandy or Sljivovica although recently; all kinds of fruit are used in spirits production. Traditional production of the fruit spirits involves the use of simple distillation pot still (Charentais alembic or French style). In the last decade, the batch distillation columns (German style) are introduced in the production of spirits in small distilleries. The distinctive identity of spirits can be affected by various distillation devices used. Very often the distillation run is left to the skills of distiller, and sometimes they have no experience to govern the distillation, resulting unpleasant and pungent flavour of the fruit spirits obtained, without distinctive identity.

For this reason, this chapter has an aim to give an overview of the basis of distillation process and main characteristics of commonly used distillation techniques in fruit spirits production.

2. Theory of distillation

Simply, distillation is the process in which a liquid is vaporised (turned to steam), recondensed (turned back into a liquid) and collected in a container. Distillation is a very old separation technology for separating liquid mixtures into their individual components by the application of heat. The basis for the component separation is differences in their boiling points. Mixture of two or more compounds is separated by heating it to a certain temperature and condensing the resulting vapours. The vapour above a boiling mixture becomes richer in more volatile components. Consequently, a boiling mixture becomes richer in less volatile components. That means that the original mixture will contain more of the less volatile material.

2.1. Distillation of binary mixture (ethanol-water)

Spirits mainly consist of the ethanol and water in quite equal portions. Alcohol has a lower boiling point than water (78.5°C compared to 100°C for water). Depending on the ratio of alcohol to water, mixture will boil at all temperatures between 78.5 and 100°C. Due to differences in boiling points, the vapour above the liquid will be richer in alcohol than water at the any moment of evaporisation. Alcohol-water relationship between gaseous and liquid alcohol-water mixtures is shown in **Figure 1**. As it can be seen in **Figure 1**, vapour phases are richer in the ethanol than the liquid, at any given mixture. Assuming, a mixture (A) of 6/94% of ethanol and water is supposed to be separated by distillation. The vapour above the liquid in the moment of equilibrium achieved will be about 42% of ethanol (A_1). Distilling the mixture of 42/58% of ethanol and water (A_1) produces a distillate that is about 78% ethanol and 22% water (A_2), and further distillation of the liquid mixture (A_2) will give a vapour with about 86% of ethanol (A_3). This means that initial alcohol strength of liquid was 6% (w/w), and after three distillations, the strength of liquid was 86% (w/w). Further, distillations will produce mixtures that are closer to the azeotropic ratio of 95.6/4.4% of ethanol and water. At this concentration, the alcohol in the vapour phase is no longer more concentrated than in the liquid phase, and fractional distillation no longer works. A mixture of this composition is called an 'azeotropic mixture'. An azeotropic mixture itself cannot be separated by ordinary distillation, and, usually, special methods are required. Generally, a third substance must be introduced into the mixture to permit separation by distillation, or some other separation schemes must be used, e.g. distillation under lower pressure not at the atmospheric pressure.

Referring to **Figure 1**, some main remarks could be given. If we put in ratio concentration of ethanol in the vapour to concentration of ethanol in the liquid (that is in this case A_1/A), we will get 42/6 = 7. It means sevenfold increase concentration of alcohol. The next step of distillation gives ratio A_2/A_1, that is 78/42 = 1.9 and finally A_3/A_2 or 86/78 = 1.1. It is obvious that the highest strength of ethanol happened when the concentration of ethanol was the lowest in the initial liquid. Here is about equilibrium ratio (*K* value) or distribution coefficient. It represents the ratio of the mole fraction some particular component in the vapour, *y*-axis to the mole fraction in the liquid *x*-axis. *K* value is defined by

$$K = \frac{y_A}{x_A} \tag{1}$$

where y_A and x_A are the mole fractions of component *A* in the vapour (*y*) and liquid phases (*x*). For the more volatile components, the *K* values are greater than 1, whereas for the less volatile components are less than 1. The *K* value is sometimes called the equilibrium 'constant', but this is misleading as it depends strongly on temperature and pressure or composition [15].

It is important to know how ease or difficult; the two components will be separated from the binary mixture (e.g. mixture of component A, more volatile component, and B less volatile component). Comparing the *K* values for these two components, relative volatility (denoted by α) obtained:

$$\alpha = \frac{\frac{yA}{xA}}{\frac{yB}{xB}} = \frac{K_A}{K_B} \tag{2}$$

Relative volatility is a measure of the differences in volatility between two components and hence their boiling points. It indicates how easy or difficult a particular separation will be. Thus, if the relative volatility between two components is equal to 1, separation is not possible by distillation. The larger the value of α, above 1, the greater the degree of separability, i.e. the easier the separation.

The values of α will be less dependent on temperature than the values of K since the K all increases with temperature in a similar manner. In general, relative volatility of a mixture changes with the mixture composition.

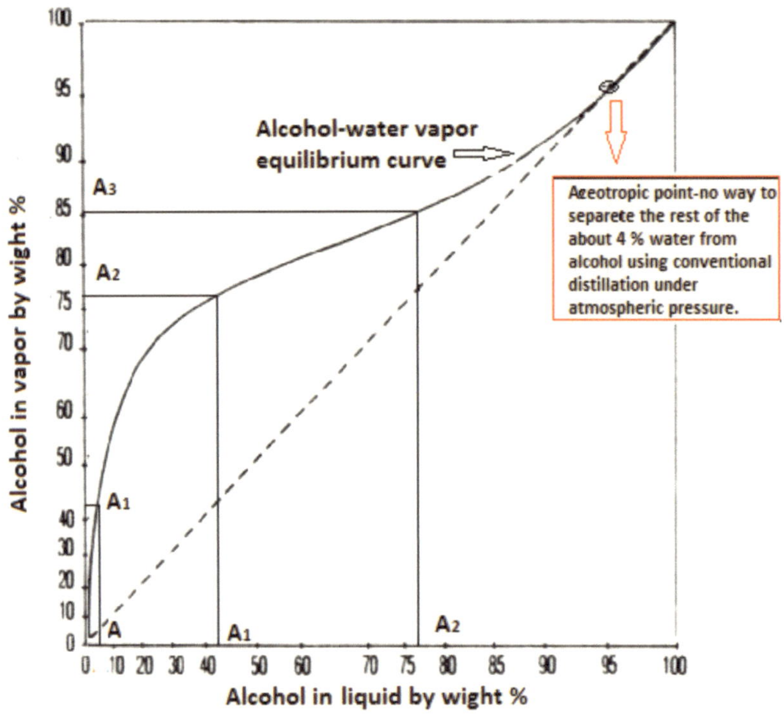

Figure 1. Vapour-liquid equilibrium of ethanol-water showing distillation steps of the mixture.

2.2. Distillation of multicomponent mixture (fermented mash)

In the production of fruit spirits, the initial material for distillation is fermented mash (or juice) that contains the ethanol and water as main compounds and a huge number of other volatile compounds that have a very large boiling point difference (e.g. acetaldehyde 20.8°C and benzaldehyde 179°C). These are collectively known as congeners, and they give the spirits authenticity and flavour. Some congeners are desirable in small quantity; others should be removed as more as possible during distillation. It is complicated to measure the relative volatility for each individual component in a multicomponent mixture. There are many reasons for that, and some of them are the following: compositions and concentrations of compounds are changing continuously with time, the aroma compounds are highly dependent on ethanol content in the liquid phase from which they are vaporising and compounds interact with themselves and between each other.

http://dx.doi.org/10.5772/66774

Distillation is performed with the aim to concentrate ethanol and desirable aroma compounds, so the relative volatility for the each individual component (i) is defined with respect to ethanol (E) and denoted by

$$\mathrm{ai} = K_i / \mathrm{K}_E \tag{3}$$

In that sense, all congeners could be separated with respect their α value on the:

- compounds with $\alpha > 1$ (these compounds are more volatile than ethanol)
- compounds with $\alpha < 1$ (these compounds are less volatile than ethanol)
- compounds with $\alpha = 1$ (separation of these compounds from ethanol is not possible)

Congeners rarely have a permanent α value greater than 1 or less than 1, due to the congeners that will distill differently depending on compositions of initial mash, their solubility in the ethanol and water, content of ethanol, variation of ethanol content in the vapour during distillation, distillation technique employed and regime of distillations. Some of the important congeners that almost ever have $\alpha > 1$ are acetaldehyde, typical representative of the first distillation fraction, or opposite, acetic acid that almost ever has $\alpha < 1$ and distills in the third fraction. Thanks to the differences in boiling points of congeners, their different solubility in the ethanol or water and the variation of ethanol content in the vapour during the distillation of all congeners will distill separately [16]. It is the basis for the separation of unwanted compounds, or concentration wanted volatile compounds during distillation of fermented fruit mash. The possibility of the separation volatile compounds enables the distiller to have a control over the process of the separation a large group of volatile compounds and ensure the production, health and pleasant quality of fruit spirits.

3. The distillation cut

During distillation ran the ethanol and water are the two main components which are actually carriers of all other volatile compounds (**Figure 2**). It could be considered that the ethanol vapour will carry over the compounds favouring spirits aroma and flavour and therefore the quality.

At the very beginning, the high volume of ethanol comes out of the still together with high volatile compounds. Through the time volume of alcohol is decreased followed by water, and low volatile compounds increased. According to this, the distillate is cut to three cuts or fractions: the head, the heart and the tail. The heads contain higher concentration of low boiling point components and mainly contain undesirable compounds. These compounds would give the distillates an unpleasant, strong and sharp flavour. In the first cut, there is a higher concentration of some toxic compounds, and therefore it must be eliminated. The best part of the run is the middle part of the distillation, the final spirits. It is a distillate rich in ethanol that is carrying a pleasant and fruity aroma compounds. The heart cut is a very clean taste lacking the sharp bite of the heads. The last cut is the tail fraction, which has to be eliminated from the heart, since it contains unpleasant fatty and oil compounds. In this fraction, the main carrier is the water. The water is carrying longer molecules, which are usually unpleasant and can be identified by the distinctive smell of 'wet dog'. The tail fractions (with or without head adding) are collected and redistilled, because it contains a relatively high concentration of alcohol and a valuable congeners.

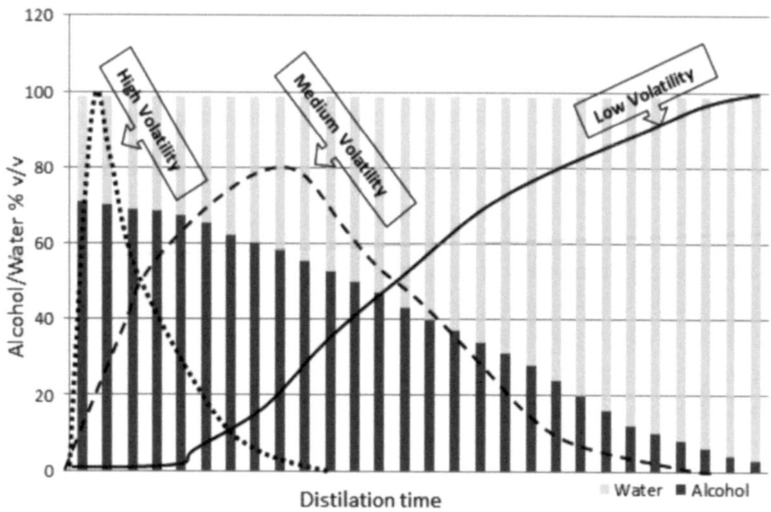

Figure 2. Distillation run of ethanol and congeners that distinguish by volatility. Adapted from © 2001 The International Centre for Brewing and Distilling [17].

3.1. How to make cuts during distillation run

In order to produce an aromatic, harmonised and pleasant fruit distillate, it is necessary to know the right time for distillation cut. During the distillation of ethanol and congeners, it is possible to manipulate the separation of volatile compounds, to clean undesirable and to concentrate desirable aroma compounds. Aroma profile of distillates very often depends on the skill of distiller to cut adequately distillation fractions. The head and tail fractions could be cut on the basis of sensory evaluation of distiller. The presence and absence of volatile congeners that give a sharp, strong and unpleasant smell to the head fraction can be cut points for the switching to the heart fraction. Also, the tail fraction starts with flavour that gives a faded, dull character to the distillates, and it should not be difficult for the sensory evaluation and separation. Experienced distillers do this very well by smell. Taste and smell still remain the most reliable method of determining when to make a cut.

The second indicator of cut points that can be used is the percent alcohol of the spirits that's flowing out of the still, especially for the separation of the heart from the tail cut. The ethanol strength could be the limiting values for the switching from the heart to the tail. This limiting value varies depending on distillation equipment involved, the fruit variety used, the quality of fermented mash, etc. Finally, the third indicator of the cut points that can be used is the temperature of the vapour before its entering to the condenser. Distiller can make the first cut in the run, when the temperature of vapour in the copper pipe reaches approximately 74–76°C. The heart cut from the tail can be made when the temperature of vapour in the copper pipe reaches around 87–88°C, and tail distills until temperature reaches 92–93°C, when distillation could be over.

Each of above-mentioned manners of distillation cut has a shortcoming, and the best way is to use all of them as guideline for the separation of congeners during distillation.

4. The two main distillation techniques

Distillation may be carried out batchwise or continuously. Nevertheless, for the production of fruit spirits, batch distillation is used. Two different types of distillation equipment are commonly used for the production of fruit spirits: copper Charentais alembic (French style) and batch distillation columns (German style). It is important to carry distillations out slowly regardless the type distillation equipment employed. Fast distillation could lead the development of hot spots and consequent accelerated thermal degradation of the final spirit.

4.1. Distillation in the Charentais alembic

In the Balkan countries, a copper alembic pot still has been utilised in artisanal distilleries, small commercial and medium-sized distilleries, in the production of fruit and wine brandy. It consists of a copper boiler, a hat, a copper pipe (pipe is not like swan neck) and a condenser (**Figure 3**). The pot still is usually heated by direct fire (fuelled by firewood or, in recent times, by natural gas). Distillation in alembic pot stills requires multiple distillations (usually double) to achieve high degree of alcohol. The aim of the first distillation is to exhaust, as much as possible, alcohol from the fermented mash. Collected distillate is called raw distillate (or low spirits) with alcohol content around 15–25% (v/v), that depends on how rich in alcohol-fermented mash was. The second stage of distillation is carried out which aimed at intensification and purification of the alcohol and has to be carried out much more carefully than the first one. In the second distillation, raw distillate is distilled with separation of three fractions. The head fraction (or cut) is collected in the amount from 1 to 2% per 100 L of raw spirits. The amount of head depends on how damaged fermented mash was. If fermented mash waits for long time until distillation is carried out, then higher head needs to be separated. The heads are thrown out because many unwanted and toxic compounds are present. Therefore, it is most important adequately to separate the heads from the hearts than the tails from the hearts. The heart fraction starts coming out at 60 to 70% (v/v) of ethanol (depends on how rich in alcohol raw spirits was) and collects until the alcohol decrease to 40–50% (v/v). The exception is Williams pear spirit. In the production of this spirit, the heart fraction needs to be cut at the lower alcohol degree (below 40%, v/v) due to the ethyl decanoate ester which distills at the beginning of the tail fraction. This ester is very important for the Williams pear spirits aroma. Research from Spaho et al. [18] showed that distillation cut of the heart fraction from the tail fraction, at 50% (v/v) of ethanol, proved to be better for the sensory impression of plum spirits made from more aromatic plum like Pozegaca and Bilska rana. The opposite rule was observed for the Stanley variety (less aromatic plum), where a better quality of brandy was achieved with a distillation cut at a lower alcohol content (40% v/v ethanol). After the heart is separated, the tail fraction distills until the end, actually, until alcohol degree achieved 3 or 5% (v/v). Measuring the alcohol content during distillation is carried out by alcoholmeter. The tail is collected and could be redistilled later or could be saved and added to the next run. During distillation, all variables are needed to keep constant and flow rate to adjust to 15–25 mL/min.

Figure 3. Traditional copper alembic pot still. The hat plays a role in the reflux because the vapour came in and will partially condense and run back down to the original liquid in the boiler and be redistilled. The flavour of final spirits was influenced by the shape and size of the hat [19].

Obtained heart fraction usually has an alcohol strength of 45–70% (v/v) that is strongly depends on kind and variety of fruit used for the spirits production [20].

Sometimes, in the Balkan countries, plum spirits, so-called Sljivovica, was produced by single-stage distillation in the alembic pot still. This type of spirits is very aromatic with so many congeners, and some of them are not desirable (**Table 1**). Sljivovica, produced by single-stage distillation, has had a double higher content of acetic acid and esters, a higher content of higher alcohols and acetaldehyde than Sljivovica produced by double distillations.

Stage of distill.	Alcohol (% v/v)	Total acids (g/L)	Aldehyde (mg/L a.a.)	Esters (mg/L a.a.)	Higher alc. (mg/L a.a.)	Methanol (% v/v a.a.)
Single	43.8	1.39	448.19	6611.4	4779.39	0.84
Double	60.6	0.36	209.68	3243.0	3264.11	1.03

Table 1. Average content (n = 5) of the main congeners in Sljivovica produced by single- and double-stage distillation [3].

Apart from ethanol, a content of methanol was higher in the Sljivovica obtained by double-stage distillation (**Table 1**). This is because the methanol follows the ethanol during distillation, and it concentrates more, together with ethanol.

4.2. Distillation in the batch distillation column

Batch distillation column style requires just one distillation to achieve high alcohol degree [21, 22]. In single-stage distillation, also these fractions are separated: the head, heart and tail. Distillation column consists of a copper pot still fitted with column with trays and a dephlegmator (**Figure 4**). In the fruit spirits production, column usually consists of three ball trays.

Figure 4. Steam-fired batch distillation column still with three trays and dephlegmator above them. Reflux rate is varied by changing the cooling rate in dephlegmator.

At the very beginning of distillation, the vapour mixture of volatile compounds goes from the boiler to the dephlegmator. Incoming vapour is partially condensed in the dephlegmator, returning a portion of it in trays. This process is called reflux, and liquid from a dephlegmator is called phlegm or reflux condensate. The reflux condensate feeds the trays in column. During distillation, the vapour coming up through the column vaporises alcohol from reflux condensate, leaving more water to keep dripping down to the next lower tray. During this countercurrent contact of vapour and liquid, which happened on each tray, the vapour stream becomes richer in light components, and the liquid stream becomes richer in heavy components. At every interface between the liquid layer and the condensed layer, contact is occurring causing greater separation of the compounds present. This process is called rectification. Consequently, as more trays are used in distillation, a greater concentration of alcohol and lower concentration of congeners are obtained. Some of these congeners are very pleasant aroma compounds, and it is not desirable to clean the alcohol too much. So, the rectification must be run very carefully, getting in mind what aroma compounds we want to have in the distillate [20, 23–25]. That means that sometimes all three trays should be active, and sometimes not, depending on how clean or how flavourful distillate needs to be.

Since the alcohol was cleaning and concentrating at the trays in the column, it starts coming out of the condenser at 70–87% (v/v). The fewer amounts of heads are collected than in alembic pot (~0.5 to 1% per 100 L of fermented mash). The head fraction is collected until content of alcohol decreased at 75–55% (v/v). The heart is obtained with approximately 65–78% (v/v) in volume around 5–10 L per 100 kg mash. After the heart is separated, the tail fraction distills until distillate obtained has an alcohol content of approximately 20–30% (v/v). All the above-mentioned alcohol contents varied from run to run and strongly depend on how much alcohol in the fermented mash has.

The main characteristic of a batch distillation column is that concentrations and temperatures are changing with time at any part of the column, so two methods can be carried out [26]: constant reflux ratio (with variable product composition at distillate) or variable reflux (with constant product composition, for one component, at distillate product). It is usually to carry out distillation with constant reflux ratio.

5. Distribution of volatile compounds during distillation by using different distillation equipment

Although both distillation techniques are performed on the same theoretical basis and in both cases three fractions have been obtained, there are several important differences in the content of ethanol and congeners that are crucial for the flavour of spirits. The first difference is content of alcohol that entails a lot of other differences. For alembic distillation, the alcoholic strength in the heart fraction is significantly lower than alcoholic strength in the heart fraction obtained by using distillation column although both distillation techniques recovered the same amount of ethanol in the hart fraction [27]. Congeners are present in very small amounts in fruit spirits but with a large influence on the bouquet and flavour. Distribution of the main congeners during distillation by using different distillation equipment, simple alembic pot and distillation column, was shown in **Figure 5.**

As it can be seen in **Figure 5**, the main differences occurred in distribution of methanol, fatty acid esters, n-propanol and isoalcohols. The main reason, for those differences, lies in the fact that those compounds show different behaviours depending on content alcohol in liquid and vapour during distillation. They will distill following their relationship with alcohol rather than their boiling point.

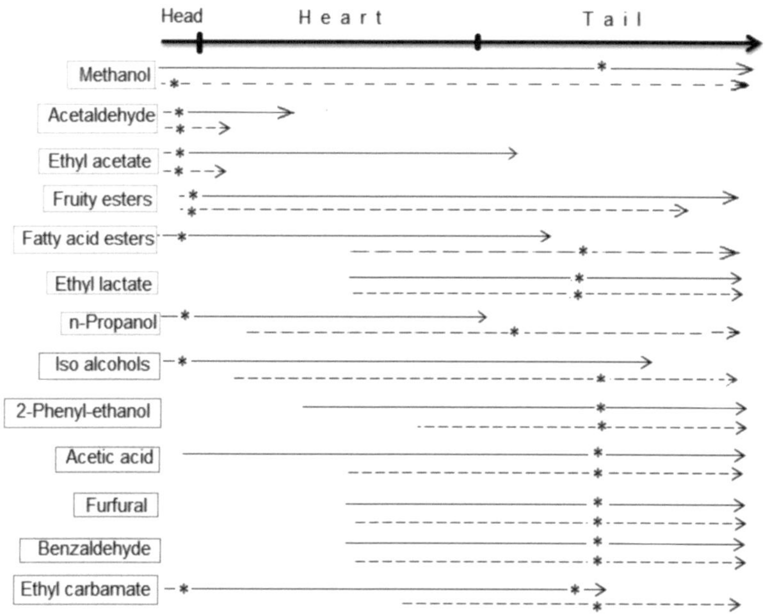

Figure 5. Distribution of main volatile compounds by using different distillation equipment: *full line*, alembic distillation; *dashed line*, column distillation and *, shows the cut where higher component is accumulated (this figure is created thanks to results of the authors) [10, 12, 18, 20, 21, 27–33].

Methanol is often the most concentrated compounds in fruit spirits [2, 8, 12, 18, 34]. Methanol is not a by-product of alcohol fermentation but is released very intensively during this process. The methanol was produced during the processing and storage of fermented mash via the effects of enzymes on pectin in the cell wall. Actually, methanol is formed from the demethoxylation of the esterified methoxyl groups in pectin.

It is characteristic to fruit brandies, significantly higher than in cereal distillates [34]. Its presence in the spirits is proof of natural origin of fruit spirits because the pectin is a natural constituent of fruits. Concentration of methanol is dependent mainly on the applied technique of the fruit treatment and the distillation and second from the fruit kind and variety.

There are different views on methanol impact to flavour of distillates. Such, Ribéreau-Gayon [35], considered the methanol imparts a cooked cabbage odour in spirits, with a threshold of 1200 mg/l. Claus and Berglund [21] wrote that methanol is considered to be a positive flavour constituent in distilled spirits. Nevertheless, most researchers say methanol is colourless volatile compounds with a mild or bland odour and does not affect the flavour of distillates [10, 32, 36, 37]. However, it is one of the most important compounds to control in the spirits due its dangerous effect to human health. In some quantities, the methanol can be dangerous because it is metabolised to formaldehyde and formic acid, which is primarily responsible for most of the toxic effects of methanol [38]. Since it is toxic to humans, the maximum level of methanol is fixed by EU Regulations No. 110/2008. According to these regulations, the concentration of methanol in fruit spirits should not exceed to 12 g/L alcohol 100% (v/v).

The boiling point of methanol is 64.7°C, and it is completely soluble in water. Considering methanol contents in the distillates obtained by different distillation techniques the results reported by several authors are vary. Methanol appears in almost equal concentration in all fractions of distillation due to the formation of azeotropic mixtures [39, 40]. It is really difficult to separate the methanol from the ethanol-water mixture. When low alcohol mixture (like fruit-fermented mash) is distilled in simple pot still, methanol will go out following his solubility in water rather than his boiling point. Methanol is highly soluble in water, therefore, methanol will distill more at the end of distillations, when vapours are richer in water. That means that methanol will accumulate more in the tail fraction [7, 32],during distillation in alembic pot still as it showed in **Figure 6**.

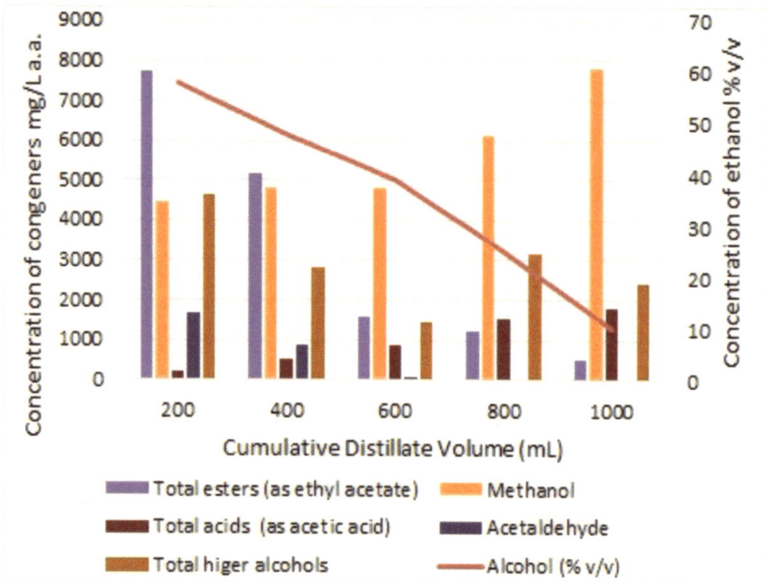

Figure 6. Concentration of the main congeners separated during distillation in alembic pot still: volume 10 L [30].

When high alcohol mixture distills, methanol will evaporate following his boiling point and will be present in the first fraction of the distillation in higher concentration. It appears mainly in the head fractions when distillation column was used [21]. Results of Cortes et al. [32] showed the concentration of the methanol was seven times higher in the case of industrial distillation (means higher concentrates and cleanses of ethanol) than the concentration of methanol in the distillates obtained by simple pot still. The opposite results are given by Arrieta-Garay [20]; there is no difference in methanol content depending on distillation system employed (alembic pot still or packed column distillations), whilst Leaute [16] and Garcia-Llobodanin et al. [27] reported that methanol content was higher in alembic distillates than in the column distillates.

The higher alcohols are quantitatively the largest group of the aroma volatile compounds in the distilled alcoholic drinks [32, 41]. They are also called fusel oil although they are not oils. It is due to form oil blotches in low alcoholic liquids because they are partially soluble in water. Higher alcohols are formed during the fermentation process and are considered as by-product of alcohol fermentation. They are produced by yeast during alcoholic fermentation, through the conversion of the branched chain amino acids present in the medium.

The chemical classes of higher alcohols include numerous alcohols. The largest share in the higher alcohol group was amylic alcohols, 2-methyl-1-butanol (active amyl alcohol), 3-methyl-1-butanol (isoamyl alcohol), 2-methyl-1-propanol (isobutyl) and 1-propanol, which are in the highest concentration present in plum brandy [18]. Very high content of 1-propanol may be an indicator for the spoilage of fruit mash.

The quantities of the other higher alcohols were very low. The shares of 1-hexanol, 1-butanol, 1-pentanol and 2-butanol were less than 5% of the total amount of the higher alcohols [18]. The compounds 1-butanol and 1-hexanol are formed during alcoholic fermentation by the hydrolysis of the corresponding acetates.

The content of 2-butanol is usually associated with a low quality of raw materials and is related to bacterial action during fermentation. Its presence negatively influences to flavour.

1-Hexanol is an alcohol originating only from raw material [2]. The level of 1-hexanol is considered the sensory relevance especially in the apple ciders and associated with a grassy scent in distillates. But, when it is present above 100 mg/L a.a. then 1-Hexanol is responsible for the very intensive grassy flavour and distillates is unpleasant both in aroma and taste. It is estimated that the presence of 1-hexanol in the above-mentioned concentrations imparts to wines and distillates fruity, liquorice and even a toothpaste-flavour profile [42].

2-Phenyl-ethanol is an aromatic alcohol and has a rose-like odour. It contributes to pleasant flavour of fruit distillates due its very low odour threshold. Bacteria, fungi and yeasts may synthesise 2-phenyl-ethanol using L-phenylalanine as a substrate, which allows this metabolite to be deemed as a potent genotypic marker for grape marc spirits [43].

2-Methyl-butanol and 3-methyl-butahnol are the most abundant minor components of spirits synthesised by yeast. In trace amounts in fruit spirits, there are also trans-3-hexenol, cis-3-hexenol and trans-2-hexenol, 1-octanol, 3-ethoxy-1-propanol allyl and benzyl, alcohol which are more characteristic for stone fruit spirits.

Higher alcohols make an important contribution to the aroma profile of distillates. They are responsible for the pleasant flavour and give an essential character of fruit distillates just when they are present in smaller quantities. However, high amounts can affect the distillate flavour, giving a strong, pungent smell and taste [2, 32]. Total higher alcohol concentrations, higher than 3500 mg/L (a.a.), are accepted as being an indicator of poor quality [43].

The level of higher alcohol in fruit spirits is influenced by fruit variety, fermentation conditions, employed distillation and distillation equipment.

Higher alcohols have boiling points lower than 200°C and are alcohol soluble and partially water soluble. During distillation of low alcohol mixture, they will distill when the vapour is rich in alcohol (they want to escape from the water in the mash, due to their low water solubility). It means they will appear mainly in the head fraction although they have high boiling point. When distilling mixture has higher concentration of alcohol (higher than 40% (v/v) [44]), higher alcohol will distill following the boiling points, and their concentration will increase as the distillation process progresses. This explains why distillates in the column showed higher concentration of higher alcohol in the tail [21] and distillates in pot still (**Figure 6**) showed higher concentration of higher alcohol in the head [18, 40].

Higher concentration of this important flavour compounds is observed in the distillates obtained in the distillation column still than in the distillates obtained in the alembic pot still [27, 40, 43].

Esters are formed during alcoholic fermentation via yeast metabolism and qualitatively present the major class of flavour compounds in distillates because they have low sensory threshold value. Esters, generally, are associated with a pleasant, fruity and flowery aroma. Their contribution to flavour is strongly influenced by their concentration.

Ethyl acetate is the major ester present in distilled alcoholic beverages. In small quantities, ethyl acetate contributes to the pleasant smell of distillates giving them a fruit character. In large amounts, it contributes to a sharp smell and gives a UHO tone (glue smell) of flavour. Ethyl acetate perception threshold of 180 g/hl a.a. gives to the spirits an acidic character [2]. High concentrations of ethyl acetate are indicative of prolonged storage of the raw material and probable acetic bacteria spoilage but could be also influenced by the distillation process. Ethyl acetate in fruit spirits constitutes even more than 80% of all the esters [18]. The importance of ethyl acetate is such that the ratio of total esters and ethyl acetate is used as indicator of quality of spirits. The higher this ratio, the higher the quality of the final product. Boiling point of ethyl acetate is 77.1°C, and it distills mainly in head fraction in both distillation techniques used. Distillates obtained in the alembic pot still showed a higher amount of ethyl acetate in comparing with distillates obtained with distillation in distillation column [16, 33], whilst some results [27] showed the higher concentration of ethyl acetate in distillate obtained by using packed distillation column.

Esters of acetic acid and higher alcohols such as isoamyl acetate, isobutyl acetate and 2-phenylethyl acetate are presence in relatively significantly amount in all fruit distillates. They are mostly responsible for the flowery and fruity aroma of the distillates. Isoamyl acetate is associated with odour and flavour of bananas, whilst isobutyl acetate is more common to raspberry flavour. Esters C_6–C_{12} are slightly higher in packed column distillates than alembic distillates [45].

Ethyl lactate is considered to give the distillates a buttery flavour and smell of rancid butter, with a perception threshold of 250 mg/l. Its presence can be linked to a malolactic fermentation, which is considered spoiled. In low concentration, lower than 154 mg/L [19], being favourable, stabilise the odour and smoothens the firm character of certain substances [2]. Ethyl lactate is associated with tail fraction of the distillate [18]. The concentration of ethyl lactate was higher in the distillates obtained by using alembic pot still than column. It comes from both the double distillation technique and the malolactic fermentation [16].

2-Phenylethyl acetate is acetate with rose odour. Although it has a high boiling point, it distills in all distilled fractions due to its partial solubility in water. Rodríguez Madrera and Mangas Alonso [10] report that 2-phenylethyl acetate distills mainly in head fractions.

http://dx.doi.org/10.5772/66774

Diethyl succinate gives a fusel-like [42] and camphor-like character. This ester derives mainly from bacterial spoilage [46] and distills in tail fraction in the bout distillation techniques.

Ethyl esters from middle-chain fatty acids (hexanoic, octanoic, decanoic and dodecanoic) are compounds of particular interest in fermented beverages and spirits on account of contributing fruity and flowery smells and their relatively high levels [47]. They are produced during the raw material fermentation. Beyond these, ethyl hexanoate is the most abundant of all middle-chain fatty acid esters. Hexanoate (ethyl caproate) supplies an aroma of fruit (banana, green apple, melon, etc.), and, so, its presence is beneficial for the spirit. Ethyl octanoate (ethyl caprylate) is more pungent and less fragrant, decanoate (caprate) is less intensive and gives fatty tones and dodecanoate (laurate) is less aromatic and has a waxy candle-like odour [19]. Despite high boiling point of fatty acid esters, they are distilled in the first fractions during pot still distillation due to their better solubility in ethanol than water [16]. During column distillation fatty acid esters are accumulated in tail fractions. There are higher amounts of these fatty acid esters in alembic distillate than in column distillate.

Ethyl esters of long-chain fatty acids are important just when they are present in higher concentration. Then, they may contribute to odours giving to spirits a candle wax and stearin tone. Esters from this group are also poorly soluble in water, so their elevated concentrations may cause turbidity and flocculation and therefore be important factors of distillate instability [48].

The major **carbonyl compound** in distillates is acetaldehyde, a direct alcoholic fermentation by-product. Moreover, significant acetaldehyde concentration may be formed by oxidation of ethanol by acetic acid bacteria in the presence of oxygen. That is one of the reasons why distillation should be carried out as soon as possible after fermentation is finished. The content of acetaldehyde is not influenced by a variety of raw material [48]; it is influenced by the strains of yeasts [49], by fermentation process and manner of distillation cut.

It has a distinctive aroma characteristic, and when it presents in higher concentration, it has a sensorial negative impact. Its importance derives not just from the pungent smell it brings along, as well as its chemical reactivity. This makes it a harmful component for the consumers. It is usually associated with intoxication and 'hangover' symptoms such as nausea, vomiting, restlessness, sweating, confusion, decrease in blood pressure, higher heart rate and headache [50]. Some authors [33, 51] established 120 g/hl a.a. of total aldehydes (mainly acetaldehyde) as limiting values before they significantly affect the aroma of the spirit.

In low concentration odour of acetaldehydes, resemble hazelnuts, cherry and overripe apples. It has a relatively low odour perception threshold. As acetaldehyde has a low boiling point, its largest concentration is distilled into the head portion of the distillate (**Figure 6**). It is completely soluble in both water and ethanol. The lower concentration of acetaldehyde is reported in distillates obtained in alembic pot still than in column distillation [16].

Acetaldehyde accounts for c. 90% (v/v) of the total aldehyde content in spirits. Other aldehydes, important for the quality of spirits, are isobutyraldehyde, 2-propenal (acrolein), and 3-hydroxy-2-butanone (acetoin) 2,3-butanedione (diacetyl). The presence of aromatic defects in spirits is related to metabolites such as acrolein (highly toxic substance) and diacetyl. Acetoin is related to carbohydrate metabolism and can be formed by enzymatic condensation of two acetaldehyde molecules or from diacetyl reduction. Acetals are products of condensation of aliphatic aldehydes and alcohols. Acetals contribute to aroma of numerous alcoholic beverages obtained from fruits. They impart a delicate pleasant taste and bouquet to alcoholic beverages. The sum of acetaldehyde + acetal, defined as total acetaldehyde, shows rather low values that confirms a mostly correct technological process, with a regular fermentation without oxidative events.

Acetaldehyde and other short-chain aliphatic aldehydes (propanal, butanal, pentanal, (E)-2-pentenal, 2-methyl-1-butanal and 3-methyl-1-butanal) have pungent, rancid and fatty odour, which may increase the tang of distilled beverages. In general, aldehydes with up to eight carbon atoms, such as acetaldehyde, formaldehyde, acrolein, benzaldehyde and furfural, have penetrating odours, generally sickening, which are considered undesirable in spirits [52]. Long-chain aldehydes are characterised by pleasant aroma. Most of them are present at levels below their individual perception thresholds [53].

Benzaldehyde is one specific aldehyde especially important for the spirits made from stone fruits like plum, cherry and apricot. Benzaldehyde is formed by hydrolysis of amygdaline present in fruit seeds and stones. It contributes to bitter almond, marzipan, cherry flavour in spirits. Benzaldehyde and benzyl alcohol are present in spirits in much higher concentrations if the mash is fermented with stones. It is not desirable in higher amount but in small amount contributes to complexity of flavour. Benzaldehyde is a high boiling point compound and distills mainly in tail regardless of distillation technique employed.

Furfural is aldehyde formed during distillation due to dehydration of residual sugars (pentoses) caused by heating in acid conditions and/or Maillard reaction. Furfural may be formed as a result of oxidation of ascorbic acid [54]. Thus, furfural is a normal constituent of fruit distillates and can be used as an indicator of distillate naturalness. Its sensor effect to the distillate can be seen through its influence to the distillate aroma that is increased by the smell of bitter almond, whilst the increased furfural concentrations may contribute to the 'hotness' of spirits.

The furfural quantity in the dried fruit distillate is naturally higher because the raw material itself has already experienced one heating process where pentose and pentosane dehydration and furfural formation occurred [55].

Furfural is a compound that is soluble in water and for this reason distills mainly in the tail fraction regardless of distillation technique employed. The higher concentration of furfural is ascribed to longer duration of distillation [56]. Its content is higher in spirits obtained by using alembic pot in comparison with the column distillation [48]. The probable reason for this is heating the alembic during double stage of distillation by direct flame.

Acetic acid accounts for more than 90% (v/v) of the total acidity in spirits. Acetic acid is a by-product of alcohol fermentation. It can be formed during the catabolism of sugar in the presence of oxygen and the yeast *Saccharomyces cerevisiae*. Acetic acid is produced by oxidation of acetaldehyde, and its content in alcoholic beverages mainly depends on the yeast strain applied. High level of acetic acid in the fermented mash is associated with contamination with acetic bacteria. In that case, the content of acetic acid is increased, and ethanol is decreased. This is owing to the conversion of the alcohol in acetic acid. A high level of acetaldehyde directly determined the acetic acid concentration in the analysed plum brandies [5].

Acetic acid is an important compound for the quality of spirits. Acetic acid has a distinctive sour taste and pungent smell, but the low concentration of acetic acid is desirable since low beverage acidity is an indicator of better quality and consumer acceptance [52]. In **Figure 6**, it can be seen that acetic acid is present in tail. It was mainly found in the last fractions [10, 16], owing to its high boiling point (117°C) although Rodríguez-Bencomo et al. [33] observed that it could be more concentrated in the head when using variable internal column reflux.

Other volatile acids important for the distillates are present in much lower quantities than acetic acid. Those are carboxylic acids and fatty acids like formic, propionic, butyric, isobutyric, caproic, undecanoic, myristic, valeric, isovaleric, 2-methylbutyric and pelargonic acids [51]. Fatty acids are known to play an important role in the sensory quality of beverages. They contribute to flavour as precursors of volatile compounds. They built esters with higher alcohols.

Caprylic, capric and lauric acids are, second to acetic acid, the most abundant free fatty acids; they are produced by yeast-mediated metabolism of carbohydrates. Short-chain free fatty acids have unpleasant odours similar to rancid butter and putrid cheese, and their presence at high levels is an indicator of poor quality fruit mash.

Hydrocyanic acid (HCN) initially occurs in bound form in the stones of the fruits and is released through enzymes during the maturation process and after the harvest. Hydrocyanic acid is formed by enzymatic hydrolysis of cyanogenic glycosides (such as amygdaline) during alcoholic fermentation. Spontaneous fermentation of fruit pulp resulted in much higher amounts of HCN in the spirits obtained in relation to the contents observed in the distillates from the mashes fermented with the addition of *S. bayanus* wine yeast. In the majority of fermented mashes, the maximum dynamics of HCN liberation was recorded on the first day of the process [57].

During hydrolysis of cyanogenic glycosides, benzaldehyde also occurred. It means that one product of hydrolysis is desirable (consumers often desire the typical 'bitter almond'), and the other be accompanied by detrimental influences and health risks [58]. It is particularly important not to largely crush the stone during preparation of stone fruits mashes for the fermentation. Another important item is that distiller should not be carried out distillation of fermented mash with stone in the pot. Investigation of Schehl et al. [58] showed that the presence or absence of stones in the mashes cannot be used as a general quality criterion. Their data provide strong evidence that the preference bitter almond flavour of spirits or the spirit, without this flavour, will remain a matter of personal taste. The boiling point of HCN is 25.7°C so it seems to be distilled in the first fraction.

Regulation (EC) No. 110 (2008) stipulates that the maximum hydrocyanic acid content in stone fruit spirits and stone fruit mark spirits shall amount to 7 g/hl of 100% (v/v) alcohol (70 mg/l).

Ethyl carbamate (EC) or urethane occurs naturally in many fermented foods and beverages. Possible routes of EC formation in alcoholic beverages are associated with the reaction between ethanol and nitrogen precursors, such as urea, carbamyl phosphate (e.g. fermented beverages) and cyanide (e.g. spirits) [59]. Cyanate is probably the ultimate precursor in most cases, reacting with ethanol to form the carbamate ester. Ethyl carbamate forms in stone fruit distillation, when exposed to light, from the natural precursors of fruit mash and ethyl alcohol. It is a potentially carcinogenic substance found in significant amounts in distilled fermented beverages, particularly in stone fruit spirits. The EC levels increase with product overheating during the distillation of spirits, especially in the distillation of some beverages rich in cyanogenic glycosides, such as amygdalin in some stone fruit brandies. Current EU legislation determines a maximum EC amount of 150 μg/l in distilled beverages.

There is high variability in ethyl carbamate content. Ethyl carbamate distills more in head and tail fractions than in the heart [59], while Alcarde et al. [60] reported that content of ethyl carbamate is increased during distillation process.

6. Alembic pot still versus column still

One of the most relevant steps in elaboration of spirits is distillation process. Distillation process can be used to correct possible mistakes that have occurred during the previous processing of the raw material. In addition, inadequate distillation can cause many defects that are difficult to eliminate by the following technological processes. During the distillation, heat facilitates the fitting of volatile compounds into resulting spirits. For the production of fruit spirits, alembic pot still and batch distillation column are the most suitable because spirits will retain the decent fruit aroma and flavour. Volatile compounds will distill differently depending on the distillation equipment used although, in accordance with results of Cortes et al. [32], for most of the compounds, there were no difference in concentration regarding the distillation equipment used. However, the concentrations of volatile compounds were influenced by processing and storage of raw material more than distillation equipment used.

Alembic stills yield better aroma that comes from fruit, the so-called primary flavour [12]. The alembic pot still produces distillates that retain the character and personality of their source ingredients. This method is slow and requires more labour, but the usage of simple copper still was preferred by several authors [16, 40]. Also, results of García-Llobodanin et al. [61] showed the distillation of pear wine with less in copper alembic leads to a better quality product.

Higher concentration of alcohol and higher separation of other volatile compounds were achieved during distillation in batch column still, giving a decent aroma of distillates. The greater yield is obtained in recovered ethanol, allowing an increased productivity by means of column distillation. This type of distillation is more effective. Nevertheless, the column-distilled spirits contained four times more esters, 20% more higher alcohols, 40% less acetaldehyde and 10% less methanol than alembic spirits [27]. Some results referred that distillates made by using a distillation column had higher sensory acceptance than distillates made in alembic pot [4, 12]. Opposite results were showed by Alcarde et al. [52]. According to these authors a sensory acceptance of cleaner fruit spirits (spirits with lower content of esters, aldehydes, higher alcohols, and methanol) was higher. The art of distillation run is to obtain the best balance between congeners present.

Also, less aromatic fruit varieties can be used to produce distillates with aromatic characteristics similar to more aromatic variety if a suitable distillation process in distillation column is used.

According to Matias-Guiu et al. [11], the traditional distillation with an alembic pot still allows limited intervention during the distillation process (only the heating power in the boiler can be manipulated) to modify the composition of the distillate. A more flexible system is the batch distillation column (in which the reflux rate can be varied in a wide range). In the same time, the other investigation [27] showed that the process with batch distillation column is much less reproducible than alembic distillation.

Claus and Berglund [21] are observed that distillates produced in the alembic pot still are usually stored in wood for many years (e.g. Cognac and Whisky), whereas the distillates produced in the distillation column are stored in the glass and consumed as clear spirits. I consider that raw material rather determinates whether distillates will be stored in wood or not. The author of this chapter considers the way of ageing that is determinates by raw material used rather than the distillation apparatus. In fruit spirits long ageing would mark the primary flavour of fruit. If distiller produced a distinctive flavour of some fruit, regardless of distillation apparatus used, it should protect the pure fruity note, not to give the distillates a strong and too much complex flavour gained during maturation in wood (quaternary flavour). If aromatic fruit spirits need to be stored in wood, then it should be just for a whilst how they would retain a fruity touch.

7. Conclusion

The choice of distillation technique using either pot still or distillation column is dependent mainly of the consumers' desire for the typical and individual flavour and style of the particular fruit spirits. In that sense, distillation method should to be adjusted for each fruit type and variety. Production of various kinds of fruit spirits requires an individual approach based on the application of knowledge and understanding the processing of desirable qualities of the spirits. Adequate managing of distillation technique enables production of the unique product.

Author details

Nermina Spaho

Address all correspondence to: n.spaho@ppf.unsa.ba

University of Sarajevo, Faculty of Agriculture and Food Sciences, Bosnia and Herzegovina

References

[1] Versini G, Franco MA, Moser S, Barchetti P, Manca G. Characterisation of apple distillates from native varieties of Sardinia island and comparison with other Italian products. Food Chemistry. 2009;**113**:1176–1183. DOI: 10.1016/j.foodchem.2008.08.003.

[2] Soufleros EH, Mygdalia AS, Natskouli P. Characterization and safety evaluation of the traditional Greek fruit distillate "Mouro" by flavor compounds and mineral analysis. Food Chemistry. 2004;**86**:625–636. DOI: 10.1016/j.foodchem.2003.11.006.

[3] Spaho N, Blesić M. The influence of distillation process on the quality of slivovica. Works of the Faculty of Agricultural and Food Sciences University of Sarajevo. 2005;**56/2**:126–135.

[4] Tesević V, Nikićević N, Jovanović A, Djoković D, Vujsić LJ, Vucković I, Bonić M. Volatile components from old plum brandies. Food Technology and Biotechnology. 2005;**43**:367–372. DOI: 10.1002/jib.62.

[5] Satora P, Tuszynski T. Influence of indigenous yeasts on the fermentation and volatile profile of plum brandies. Food Microbiology. 2010;**27**:418–424. DOI: 10.1016/j.fm.2009.12.005.

[6] McGavin J. German Fruit Schnaps and Brandies [Internet]. 2016. Available from: http://germanfood.about.com/od/drinks/tp/German-Fruit-Schnaps.htm [Accessed: 2016-06-03].

[7] Hernandez-Gomez LF, Ubeda-Iranzo J, Garcıa-Romero E, Briones-Perez A. Comparative production of different melon distillates: chemical and sensory analyses. Food Chemistry. 2005;**90**:115–125. DOI: 10.1016/j.foodchem.2004.03.033.

[8] Ruso Coldea TE, Socaicu C, Dan Vodnar MP. Gas-chromatographic analysis of major volatile compounds found in traditional fruit brandies from Transylvania, Romania. Notulae Botanicae Horti Agrobotanici Cluj-Napoca. 2011;**39**:109–116. DOI: 10.15835/nbha3926053.

[9] Biernacka P, Wardencki W. Volatile composition of raw spirits of different botanical origin. Journal of the Institute of Brewing. 2012;**118**:393–400. DOI: 10.1002/jib.55.

[10] Rodríguez Madrera R, Mangas Alonso JJ. Distribution of the principal minor volatiles during cider distillation in 'alquitara'. Acta Alimentaria. 2010; **40**:262–269. DOI: 10.1556/AAlim.2010.0005.

[11] Matias-Guiu P, Rodríguez-Bencomo JJ, Orriols I, Pérez-Correa JR, López F. Floral aroma improvement of Muscat spirits by packed column distillation with variable internal reflux. Food Chemistry. 2016;**213**:40–48. DOI: 10.1016/j.foodchem.2016.06.054.

[12] Arrieta-Garay Y, García-Llobodanin L, Pérez-Correa JR, López-Vázquez C, Orriols I, López F. Aromatically enhanced pear distillates from Blanquilla and conference varieties using a packed column. Journal of Agricultural and Food Chemistry. 2013;**61**:4936–4942. DOI: 10.1021/jf304619e.

[13] Rodriguez Madrera R, Blanco Gomis D, Mangas Alonso JJ. Characterization of cider brandy on the basis of aging time. Journal of Food Science. 2003;**68**:1958–1961.

[14] Profumo A, Riolo C, Pesavento M, Francoli A. Evolution of the Italian distillate "grappa" during aging in wood: a gas chromatographic and high performance liquid chromatographic study. American Journal of Enology and Viticulture. 1988;**39**:273–278.

[15] Halvorsen IJ, Skogestad S. (Distillation Theory) [Internet]. Available from: /home/ivarh/thesis/book/DistillationTheory_ch.fm Version: 11 August 2000 [Accessed: 2016-06-04].

[16] Léauté R. Distillation in alambic. American Journal of Enology and Viticulture. 1990;**41**:90–103.

[17] The International Centre for Brewing and Distilling 2001 [Internet]. Available from: http://cocktailchem.blogspot.ba/2014/06/the-physics-of-pot-stills.html- [Accessed: 2016-06-04].

[18] Spaho N, Dürr P, Grba S, Velagić-Habul E, Blesić M. Effects of distillation cut on the distribution of higher alcohols and esters in brandy produced from three plum varieties. Journal of the Institute of Brewing. 2013;**19**:48–56. DOI: 10.1002/jib.62.

[19] Bougas NV. Evaluation the Effect of Pot Still Design on the Resultant Distillate. [Master thesis]. Stellenbosch University: Faculty of AgriScience; 2009.

[20] Arrieta-Garay Y. Influence of a Packed Distillation Column on Volatile Composition and Sensory Profile of Spirit Drinks: Application to Pear, Kiwi and Grape Pomace. [Doctoral thesis]. Departament d'Enginyeria Química Facultat d'Enologia, Universitat Rovira i Virgili, Tarragona; 2014.

[21] Claus MJ, Berglund KA. Fruit brandy production by batch column distillation with reflux. Journal of Food Process Engineering. 2005;**28**:53–67. DOI: 10.1111/j.1745-4530.2005.00377.x.

[22] Durr P. Brennen. In: Durr P, Albercht W, Gossinger M, Hagmann K, Pulver D, Scholten G. Technologie der Obstbrennerei. 3. Volling auflage. Eugen Ulmer; 2010. pp. 215–241.

[23] Gaymon JF, Chemical aspects of distilling wines into brandy. In: Dinsmoor Webb A, editor. Chemistry of Winemaking. American Chemical Society; 1974. pp. 232–253. Available from: https://www.stilldragon.org/uploads/.../37/54608215f4a2e07a122181675ef4ab.pdf - [Accessed: 2016-06-04].

[24] Osorio D, Perez-Correa JR, Biegler LT, Agosin E. Wine distillates: practical operating recipe formulation for stills. Journal of Agriculture and Food Chemistry. 2005;**53**;6326–6331. DOI: 10.1021/jf047788f.

[25] Sliwinska M, Wisniewska P, Dymerski T, Wardencki W, Namiesnik J. The flavour of fruit spirits and fruit liqueurs: a review. Flavour and Fragrance Journal. 2015;**30**:197–207. DOI: 10.1002/ffj.3237.

[26] Narvaez-Garcia A, Zavala-Loria JC, Vilchis-Bravo LE, Rocha-Uribe JA. Design of batch distillation columns using short-cut method at constant reflux. Journal of Engineering. 2013;Article ID 685969:1–14. http://dx.doi.org/10.1155/2013/685969.

[27] Garcıa-Llobodanin L, Roca J, Lopez JR, Perez-Correa JR, Lopez F. The lack of reproducibility of different distillation techniques and its impact on pear spirit composition. International Journal of Food Science and Technology. 2011;**46**:1956–1963. DOI: 10.1111/j.1365-2621.2011.02707.x.

[28] Scholten G. Untersuchungsmethoden und Betriebskontrolle In: Durr P, Albercht W, Gossinger M, Hagmann K, Pulver D, Scholten G. Technologie der Obstbrennerei. 3. Volling auflage. Eugen Ulmer; 2010. 275 p.

[29] Spaho N, Alihodžić A, Begić-Akagić A, Blesić M. Content of methanol in the apple pomace distillates. Works of the Faculty of Agricultural and Food Sciences University of Sarajevo. 2010;**LV**:201–211.

[30] Faletar J, Blesic M, Smajic M, Begic-Akagic A, Alihodzic A, Spaho N. Dynamics of evaporation of the certain volatiles during plum brandy distillation. In: Proceedings of the 24th International Scientific-Expert-Conference of Agriculture and Food Industry; 25–28 September 2013; Sarajevo, Bosnia and Herzegovina. pp. 204–208.

[31] Riachi LG, Santos Â, Moreira RFA, De Maria CAB. A review of ethyl carbamate and polycyclic aromatic hydrocarbon contamination risk in cachaça and other Brazilian sugarcane spirits. Food Chemistry. 2014;**149**:159–169. http://dx.doi.org/10.1016/j.foodchem.2013.10.088.

[32] Cortes S, Gil ML, Fernandez E. Volatile composition of traditional and industrial Orujo spirits. Food Control. 2005;**16**:383–388. DOI: 10.1016/j.foodcont.2004.04.003.

[33] Rodríguez-Bencomo JJ, Pérez-Correa JR, Orriols I, López F. Spirit distillation strategies for aroma improvement using column reflux. Food and Bioprocess Technology. 2016. DOI: 10.1007/s11947-016-1776-0.

[34] Coldea TA, Socaciu C, Tofana M, Vekony E, Ranta N. Impact of distillation process on the major volatile compounds as determined by GC-FID analysis in apple bandy originated from Transylvania, Romania. Bulletin UASVM Agriculture. 2012;**69**:228–235.

[35] Ribéreau-Gayon P, Glories Y, Maujean A, Dubordieu D. The chemistry of wine, stabilisation and treatments. In: Handbook of Enology Vol. **2**, 2nd ed. John Wiley and Sons, New York; 2000. pp. 41–54.

[36] Gueven A. Chemical fingerprints of Raki: a traditional distilled alcoholic beverage. Journal of the Institute of Brewing. 2013;**119**:126–132. DOI: 10.1002/jib.75.

[37] Geroyiannaki M, Komaitis ME, Stavrakas DE, Polysiou M, Athanasopoulos PE, Spanos M. Evaluation of acetaldehyde and methanol in greek traditional alcoholic beverages from varietal fermented grape pomaces (*Vitis vinifera* L.). Food Control. 2007;**18**:988–995. DOI: 10.1016/j.foodcont.2006.06.005.

[38] Kraut JA, Kurtz I. Toxic alcohol ingestions: clinical features, diagnosis, and management. Clinical Journal of the American Society of Nephrology. 2008;**3**:208–225. DOI: 10.2215/CJN.03220807.

[39] Spaho N. Effects of distillation cut in plum raw distilates on distribution of higher alcohols and esters. [Doctoral thesis]. Library of Faculty of Agriculture and Food Sceince, University of Sarajevo; 2007.

[40] Hernandez-Gomez LF, Ubeda J, Brions A. Melon fruit distillates: comparison of different distillation methods. Food Chemistry. 2003;**82**:539–543.

[41] Tanamool V, Leelavatcharamas V, Moon-amart S, Vichitphan K. Major volatile compounds of fruit distillates prior to aging for fruit brandy. In: Proceeding of 1st International Conference on Fermentation Technology for Value Added Agricultural Products. Fermentation Technology for Value Added Agricultural Products (FerVAAP), Khon Kaen University, Khon Kaen, Thailand; 2005.

[42] Ferreira V, Hernandez-Orte P, Escudero A, Lopez R, Cacho J. Semipreperative reversed-phase liquid chromatographic fractionation of aroma extracts from wine and other alcoholic beverages. Journal of Chromotography A. 1999;**864**:77–88. DOI: 10.1016/S0021-9673(99)01004-3.

[43] Cortés S, Rodríguez R, Salgado JM, Domínguez JM. Comparative study between Italian and Spanish grape marc spirits in terms of major volatile compounds. Food Control. 2011;**22**:673–680.

[44] Panker RJ, Boucher AR. Continuous distillation. In: Piggott JR, Sharp R, Duncan REB, editors. The Science and Technology of Whiskies. Longman Scientific and Technical; 1989. pp. 150–181.

[45] Arrieta-Garay Y, López-Vázquez C, Blanco P, Pérez-Correa JR, Orriols I, López F. Kiwi spirits with stronger floral and fruity characters were obtained with a packed column distillation system. Journal of the Institute of Brewing. 2014;**120**:111–118. DOI: 10.1002/jib.117.

[46] Karagiannis S, Lanaridis P. Insoluble grape material present in must affects the overall fermentation aroma of dry white wines made from three grape cultivars cultivated in Greece. Journal of Food Science. 2002;**67**:369–374. DOI: 10.1111/j.1365-2621.2002.tb11412.x.

[47] Rodríguez Madrera R, García Hevia A, Valles BS. Comparative study of two aging systems for cider brandy making. Changes in chemical composition. LWT—Food Science and Technology. 2013;**54**:513–520. http://dx.doi.org/10.1016/j.lwt.2013.05.037.

[48] Lukić I, Tomas S, Milicević B, Radeka S, Perusić Đ. Behaviour of volatile compounds during traditional alembic distillation of fermented Muscat Blanc and Muskat Ruza Porecki grape marcs. Journal of the Institute of Brewing. 2011;**117**:440–450. DOI: 10.1002/j.2050-0416.2011.tb00491.x.

[49] Kwak HS, Seo JS, Hur Y, Shim HS, Lee Y, Kim M, Jeong Y. Influence of yeast strains on the physicochemical characteristics, methanol and acetaldehyde profiles and volatile compounds for Korean rice distilled spirit. Journal of the Institute of Brewing. 2015;**121**:574–580. DOI: 10.1002/jib.268.

[50] Masson J, Cardoso MG, Zacaroni LM,, Anjos JP, Sackz AA, Machado AMR, Nelson DL.Determination of acrolein, ethanol, volatile acidity, and copper in different samples

of sugarcane spirits. Ciência e Tecnologia de Alimentos, Campinas. 2012;**32**:568–572. DOI: 10.1590/S0101206120120050000075.

[51] Silva ML, Macedo AC, Malcata FX. Review Steam distilled spirits from fermented grape pomace. Food Science and Technology International. 2000;**6**:285–300. DOI: 10.1177/108201320000600403.

[52] Alcarde AR, Souza PA, Belluco AES. Chemical profile of sugarcane spirits produced by double distillation methodologies in rectifying still. Ciência e Tecnologia de Alimentos, Campinas. 2011;**31**:355–360. ISSN 0101-2061.

[53] López-Vázquez C, Orriols I, Perelló MC, Revel G. Determination of aldehydes as pentafluorobenzyl derivatives in grape pomace distillates by HS-SPME-GC/MS. Food Chemistry. 2012;**130**:1127–1133. DOI: 10.1016/j.foodchem.2011.07.140.

[54] Briones A, Ubeda-Iranzo J, Hernández-Gómez L. Spirits and Liqueurs from melon fruits (*Cucumis melo* L.). In: Zereshki S, editors. Distillation—Advances from Modeling to Applications. InTech; 2012. [Internet] Available from: http://www.intechopen.com/books/distillation-advances-from-modelingto-applications/spirits-and-liqueurs-from-melon-fruits [Accessed: 2014.08.04].

[55] Spaho N, Blesić M, Begić-Akagić A, Oručević S, Pjano A, Ferizović H. Furfural content in the distillates of prune and dried pear. In: Proceedings of the Joint Central European Congress; 15–17 May 2008; Cavtat, Croatia; pp.143–149.

[56] Jasarevic E, Sarenac Z, Spaho N. The effect ot the temperature of heating during distillation on the content of furfural and aroma of apple destillates. In: Book of Abstract; 1st Students Congress "Food-Nutrition-Helth". 7–9 July 2016; Sarajevo; 4 p.

[57] Balcerek M, Szopa J. Ethanol biosynthesis and hydrocyanic acid liberation during fruit mashes fermentation. Czech Journal of Food Science. 2012;**30**:144–152.

[58] Schehl B, Lachenmeier D, Senn T, Heinsch JJ. Effect of the stone content on the quality of plum and cherry spirits produced from mash fermentations with commercial and laboratory yeast strains. Journal of Agricultural and Food Chemistry. 2005;**53**:8230–8238. DOI: 10.1021/jf0511392.

[59] Riachi LG, Santos Â, Moreira RFA, C.A.B. De MariaA review of ethyl carbamate and polycyclic aromatic hydrocarbon contamination risk in cachaça and other Brazilian sugarcane spirits. Food Chemistry. 2014;149:159–169.

[60] Alcarde AR, Souza LM, Bortoletto AM. Ethyl carbamate kinetics in double distillation of sugar cane spirit. Journal of the Institute of Brewing. 2012;**118**:352–355. DOI: 10.1002/jib.48.

[61] García-Llobodanin L, Achaerandio I, Ferrando M, Güell C, López F. Pear Distillates from pear juice concentrate: effect of lees in the aromatic composition. Journal of Agricultural and Food Chemistry. 2007;**55**:3462–3468. DOI: 10.1021/jf0633589.

Chapter 7

Fractional Distillation of Organic Liquid Compounds Produced by Catalytic Cracking of Fats, Oils, and Grease

C. C. Fereira, E. C. Costa, D. A. R. de Castro,
M. S. Pereira, A. A. Mâncio, M. C. Santos,
D. E. L. Lhamas, S. A. P. da Mota, M. E. Araújo,
Luiz E. P. Borges and N. T. Machado

Additional information is available at the end of the chapter

http://dx.doi.org/10.5772/66759

Abstract

This work aims to investigate the fractional distillation of organic liquid products (OLP) obtained by catalytic cracking of palm oil (*Elaeis guineensis* Jacq.) at 450°C, 1.0 atm, with 5, 10, and 15% (wt) Na_2CO_3, using a stirred tank reactor of 143 L. The fractional distillations of OLP were carried out in laboratory scale with and without reflux using columns of different heights, and a pilot-packed distillation column with internal reflux. OLP and distillation fractions (gasoline, kerosene, light diesel, and heavy diesel) were physicochemically characterized for density, kinematic viscosity, acid value, saponification value, refractive index, flash point, and copper strip corrosion. The OLP and light diesel fractions were analyzed by Fourier transform infrared spectroscopy (FT-IR) and gas chromatography-mass spectrometry (GC-MS). For the experiments in laboratory scale, the yields of distillates decrease along with column height, with and without reflux, while those of bottoms products increase. The yields of distillates and gas increase with increasing Na_2CO_3 content, while those of bottoms products decrease. The densities of gasoline, kerosene, and light diesel produced in laboratory scale with reflux superpose exactly those of kerosene, light diesel, and heavy diesel produced in laboratory scale without reflux. The kinematic viscosity decreases with increasing column height for the experiments in laboratory scale. The acid values of distillation fractions decrease along with the column height for the experiments with and without reflux. The FT-IR of distillation fractions in pilot and laboratory scales identified the presence of aliphatic hydrocarbons and oxygenates. The GC-MS analysis identified OLP composition of 92.84% (area) hydrocarbons and 7.16% (area) oxygenates. The light diesel fraction contains 100% hydrocarbons with an acid value of 0.34 mg KOH/g, proving the technical feasibility of OLP de-acidification by the fractional distillation process.

Keywords: palm oil, organic liquid products, fractional distillation, light diesel

1. Introduction

Pyrolysis and/or catalytic cracking is one of the most promising processes to convert triacylglycerides (TAGs), the major compounds of vegetable oils and animal fats [1, 2], into liquid biofuels [3], and the literature reports several studies on the subject [3–47]. Both processes have the objective of obtaining hydrocarbons for use as fuels [3, 4, 6–24, 28–37]. However, the chemical composition of organic liquid products (OLP) shows a significant difference because of the complex cracking mechanism of TAGs [4, 5, 10, 19, 25–27]. Besides the type of cracking mode (thermal cracking and thermal catalytic cracking), other factors that significantly affect the liquid fuel composition are the characteristics of raw material, reaction temperature, residence time, mode of operation (fluidized bed reactor, sludge bed reactor, etc.), and the presence of water in the raw material and/or in the catalyst [6–11, 14, 15, 19, 21–24, 28–30].

The reaction products obtained by pyrolysis and/or catalytic cracking of oils, fats, grease, and fatty acid mixtures include gaseous and liquid fuels, water, and coke [6–8, 14, 15, 17, 21–24, 28–30]. The physicochemical properties and chemical composition of OLP depend on the selectivity of the catalyst used [6, 7, 10, 14–17, 20, 21–25, 28–39]. The OLP consists of hydrocarbons [8, 11, 12, 16, 17, 21–24, 26, 27], corresponding to the boiling point range of gasoline, kerosene, diesel fossil fuels, and oxygenates [6–8, 11, 12, 15–17, 21–25].

One of the advantages of catalytic cracking of oils, fats, greases, and fatty acid mixtures is the possibility of using low-quality lipid-based materials [6, 7, 20, 21–24, 28–35, 41] and the compositional similarities of OLP to fossil fuels [3, 6–8, 10, 21–24]. The OLP obtained by catalytic cracking presents lower amounts of carboxylic acids compared to pyrolysis, because of the catalytic activity in the secondary cracking step, where the carboxylic acids are broken up to form hydrocarbons [10], as reported elsewhere [21–23, 30]. The OLP can not only be stored and transported, but can also be refined and/or upgraded by applying physical (filtration, decantation, and centrifugation) and thermal separation processes (distillation, liquid-liquid extraction, and adsorption) to produce high-quality green fuel-like fractions with the potential to substitute partially fossil fuels [6, 11, 16, 21–23, 40, 44].

The disadvantages and/or drawbacks of OLP obtained by pyrolysis and/or catalytic cracking of oils, fats, greases, and fatty acid mixtures are the high acid value [8, 11, 14, 19, 22, 45, 46] and high concentrations of olefins, making OLP a corrosive and unstable fuel [9, 21]. To increase the yield of OLP and reduce undesired reaction products, as well as the content of oxygenate compounds, a wide variety of catalysts have been tested in catalytic cracking, particularly zeolites [6, 7, 10, 12, 13, 15–18, 21, 22, 28–39]. However, OLP obtained by catalytic cracking using zeolites and mesoporous catalysts still has a high carboxylic acid content [7, 14, 15, 45].

In this context, studies have been investigating strategies to minimize the high acid values and high concentration of olefins in OLP obtained by catalytic cracking of oils, fats, greases, and fatty acid mixtures, including the application of cheap alkali catalysts such as Na_2CO_3 to reduce the acid value of liquid biofuels [7, 21, 23, 24, 30, 47, 51–55]. OLP with lower acid values makes it possible to apply physical (filtration, decantation, sedimentation, and centrifugation) [53–55], chemical (neutralization) [53–55], and thermal separation processes (distillation, liquid-liquid extraction, and adsorption) to produce high-quality green hydrocarbon-like fuels [21–24, 44, 53–55]. In the last few years, processes have been proposed to remove and/or recover oxygenate compounds from biomass-derived bio-oils including molecular distillation to separate water and carboxylic acids from pyrolysis bio-oils [56–58], fractional distillation to isolate/enrich chemicals and improve the quality of bio-oil [59–64], and liquid-liquid extraction using organic solvents and water to recover oxygenate compounds of bio-oils [40, 65]. Non-conventional separation methods using aqueous salt solutions for phase separation of bio-oils are also applied [66]. Furthermore, the literature reports several studies upon fractionation of OLP by single-stage and multistage distillation to obtain hydrocarbon-like fuels in the temperature boiling point range of gasoline, kerosene, and diesel-like fractions [6, 7, 11, 14, 15, 21–24, 30, 35, 37, 39, 41, 46–55]. However, until now only a few studies have investigated systematically the effect of column height on the chemical composition of OLP [6, 7, 47], but no systematic study has investigated the effect of column height, reflux ratio, and OLP composition on the physicochemical properties of the distillation fraction of OLP.

This work aims to investigate the effect of column height, reflux rate, and OLP composition on the physicochemical properties of distillation fractions and de-acidification of OLP by fractional distillation using laboratory columns of different heights and a pilot-packed distillation column with internal reflux.

2. Materials and methods

2.1. Materials

OLP was obtained by catalytic cracking of crude palm oil (*Elaeis guineensis* Jacq.) at 450°C, 1.0 atm, with 15% (wt) Na_2CO_3 in a stirred tank slurry reactor of 143 L, operating in batch mode, as described in a similar study reported elsewhere [21].

2.2. Physicochemical analysis of palm oil, OLP, and distillation fractions

Palm oil, OLP, and distillation fractions have been physicochemically characterized for acid value (AOCS Cd 3d-63), saponification value (AOCS Cd 3-25), free fatty acid content (ASTM D5555), density (ASTM D1480) at 25°C, kinematic viscosity (ASTM D445/D446), flash point (ASTM D93), copper strip corrosion (ABNT/NBR 14359), and refractive index (AOCS Cc 7-25).

2.3. Fractional distillation of OLP

2.3.1. *Laboratory unit*

2.3.1.1. *Distillation without reflux: experimental apparatus and procedures*

The laboratory fractional distillation apparatus was operated without reflux and the procedure is described in detail elsewhere [21].

2.3.1.2. Distillation with reflux: experimental apparatus and procedures

The fractional distillation of OLP with reflux was performed by using an experimental apparatus similar to that described by Mota et al. [21]. The distillation apparatus had a thermostatically controlled electrical heating blanket of 480 W (Fisaton, Model: 202E, Class: 300), and a 500 mL round bottom, three neck borosilicate-glass flask with outer joints, and side joints angled at 20°, 24/40. The side joints used to insert a long thin thermocouple of a digital thermometer and the other used to collect samples, the center joint, 24/40, were connected to a distillation column (Vigreux) of different heights (L_1 = 10 cm, L_2 = 30 cm, L_3 = 50 cm). The borosilicate-glass distillation columns (Vigreux) with bottom inner and top outer joints 24/40 were connected to an inverted Y-type glass support, the left side bottom inner joint 24/40 was connected to the distillation column top outer joint 24/40, and the right side bottom inner joint 24/40was connected to the 250 mL glass separator funnel top outer joint 24/40. The center top outer joint 24/40 was connected to the bottom inner joint 24/40 of a Liebig glass-borosilicate condenser. The right side of the inverted Y-type glass support had a Teflon valve that made it possible to drip only a part and/or fraction of liquid condensates into the glass separator funnel, thus creating a reflux rate. A thermocouple connected to the top outer joint 7/25 of the left side of the inverted Y-type glass support made it possible to measure the vapor temperature at the top of the borosilicate-glass distillation columns (Vigreux). A cryostat bath (VWR Scientific, Model/Series: 913174) provided cold water at 15°C to the Liebig glass-borosilicate condenser. The 500 mL round bottom borosilicate-glass flask and the distillation column (Vigreux) were insulated with glass wool and aluminum foil sheet to avoid heat losses. Initially, approximately 300 g of OLP was weighed, the heating system was switched on, and the distillation time and temperature were recorded. From the time the vapor phase started to condensate, the Teflon valve was regulated to a reflux rate of two drops per second. The mass of distillation fractions (gasoline, kerosene, light and heavy diesel-like fuels) was recorded and weighed. The distillation fractions were submitted to the pretreatment of decantation to separate the aqueous and organic (OLP) phases.

2.3.2. Distillation pilot unit

Figure 1 illustrates the fractional distillation unit (Goel Scientific Glass Works Pvt. Ltd, India, Model: FDU50) with dimensions (height = 370 cm, length = 90 cm, depth = 60 cm), operating pressure –1.0–0.5 bar for process, utility, and vessel sides, and maximum operating temperature of –50–300°C, constructed of borosilicate glass and 100% polytetrafluoroethylene (PTFE). The distillation unit consisted of a distillation vessel of 50 L with a drain valve (DN 25), a heating/cooling system of 6.0 kW, and a heating surface of 0.5 m^2 (heating medium), as well as copper coils of 0.4 m^2 heating the transfer area and an operating pressure up to 10.0 bar (steam). A digital display controlled the heating rate and distillation vessel temperature and displayed the temperature at the reflux divider, operating range 0–300°C, ±2.0°C tolerance, PT-100 sensor for the distillation vessel (in built), and reflux divider. The vapor line (H = 100 cm, DN 100) was packed with cylindrical borosilicate-glass raschig rings of 15 mm length and 10 mm diameter, and the vapor condenser (DN 100), cooled with water, had a heating transfer surface of 0.5 m^2 with a manually operated reflux divider. The product cooler had a 0.2 m^2 heating transfer area, coupled to two twin receivers of 5 and 10 L with spherical geometry, the 10 L spherical vessel with a drain valve

http://dx.doi.org/10.5772/66759

(a) (b)

Figure 1. Differential glass packed distillation unit of 50 L: (a) Frontal view, (b) Lateral view.

(DN 25). The 50 L round borosilicate-glass vessel and the distillation column were insulated with glass wool and aluminum foil sheet to avoid heat losses. Initially, approximately 9.50 kg of OLP was weighed and introduced inside the distillation vessel and the electrical heating system switched on for a heating rate of 2°C/min, being the distillation time and temperature recorded. Afterwards, the freshwater cooling system valve was opened. From the time the vapor phase started to condensate, the regulating valve between the reflux divider and the product cooler was open. The mass of distillation fractions (gasoline, kerosene, and light diesel-like fuels) was recorded and weighed. The distillation fractions were submitted to the pretreatment of decantation to separate the aqueous and organic (OLP) phases.

2.4. Chemical analysis of OLP and distillation fractions

2.4.1. Physicochemical analysis of distillation fractions

Distillation fractions (gasoline-like fraction: 40°C < T^B < 175°C; kerosene-like fraction: 175°C < T^B < 235°C; light diesel-like fraction: 235°C < T^B < 305°C; and heavy diesel-like fraction: 305°C < T^B < 400°C) were characterized according to the analysis described in Section 2.2, except for flash point and free fatty acid content. FT-IR of OLP and distillation fractions (gasoline: 40°C < T^B < 175°C; kerosene: 175°C < T^B < 235°C; light diesel: 235°C < T^B < 305°C; and heavy diesel: 305°C < T^B < 400°C) were performed using an FT-IR spectrometer as described in detail elsewhere [21, 22]. Prior to the chemical analysis by GC-MS, described in detail elsewhere [22, 23], the samples of OLP and light diesel-like fraction (235°C < T^B < 3055°C) were submitted to a pretreatment of chemical derivatization of free fatty acids.

3. Results and discussions

3.1. CPO and OLP physicochemical properties

Table 1 illustrates the physicochemical characterization of crude palm oil (*Elaeis guineensis* Jacq.) and OLP obtained by catalytic cracking of palm oil at 450°C and 1.0 atm, with 5, 10, and 15% (wt) Na_2CO_3 in pilot scale. Crude palm oil (CPO) used as raw renewable material on the catalytic cracking experiment was physicochemically characterized in a previous study [21].

3.2. Catalytic cracking of CPO

The process conditions, material balance, and yields of reaction products (OLP, coke, gas, and H_2O) obtained by catalytic cracking of CPO at 450°C and 1.0 atm, with 15% (wt) Na_2CO_3, are shown in **Table 2**. The obtained OLP yield was lower, but in accordance with similar studies reported in the literature [21–24, 30]. The gas yield was lower than that reported in similar studies [21–24], while the yield of coke was higher, but in accordance with that reported elsewhere [21–24, 30].

Physicochemical properties	Palm oil	OLP			ANP No. 65
		(wt%) Na_2CO_3			
	(21)	5	10	15	
ρ (g/cm³)	0.900	0.849	0.834	0.830	0.82–0.85
Acid value (mg KOH/g)	4.80	51.56	4.10	3.55	–
$\frac{(IA_{PalmOil} - IA_{OLP})}{IA_{PalmOil}} * 100\ (\%)$	–	–	14.58	26.04	–
Refractive index (–)	1.460	1.458	1.458	1.454	–
μ (cSt)	48.05	6.59	5.67	4.82	2.0–4.5
Flash point (°C)	–	87	28	27	>38
Saponification value (mg KOH/g)	179.90	70.95	64.91	54.15	–
$\frac{(IS_{PalmOli} - IS_{OLP})}{IS_{PalmOil}} * 100\ (\%)$	–	60.56	63.92	69.90	–
Ester index (mg KOH/g)	174.60	19.39	60.81	50.60	–
Content of FFA (%)	2.40	25.78	2.05	1.78	–
Copper strip corrosion (IA)	–	1A	1A	1A	1A

ANP: Brazilian National Petroleum Agency, Resolution No. 65 (Specification of Diesel S10).
IA, acid value; IS, saponification value, Ester index, IS – IA, Free Fatty Acids (FFA).

Table 1. Physicochemical analysis of palm oil and OLP obtained by catalytic cracking of palm oil at 450°C and 1.0 atm, with 5, 10, and 15% (wt) Na_2CO_3 in pilot scale.

Process parameters	Na_2CO_3 15% (wt)
	Pilot
Cracking temperature (°C)	450
Mass of palm oil (kg)	34.90
Mass of Na_2CO_3 (g)	5.24
Cracking time (min)	100
Mechanical stirrer speed (rpm)	150
Initial cracking temperature (°C)	320
Yield of OLP (wt%)	58.74
Yield of coke (wt%)	15.47
Yield of H_2O (wt%)	13.64
Yield of gas (wt%)	12.15

Table 2. Process parameters and overall steady-state material balance of catalytic cracking of palm oil at 450°C, 1.0 atm, with 15% (wt) Na_2CO_3 in pilot scale.

3.3. Fractional distillation of OLP

3.3.1. Laboratory unit

3.3.1.1. Distillation without reflux: material balances and yields of distillation fractions

Table 3 illustrates the material balances and yields of distillation products (distillates, bottoms, and gas) produced by laboratory fractional distillation of OLP obtained at 450°C and 1.0 atm, with 5, 10, and 15% (wt) Na_2CO_3 in pilot scale, using Vigreux columns of different heights (L_1 = 10 cm, L_2 = 30 cm, L_3 = 50 cm), operating without reflux. For the experiments carried out using columns of different heights, with and without reflux, the yields of distillates (biofuels) and gas decreased in a smooth exponential and linear fashion, respectively, along with the column height, while that of bottoms products increased exponentially with increasing column height, as shown in **Figure 2**. The same tendency was observed by Dandik and Aksoy [6, 7]. The yield of distillates of 66.26% (wt), obtained with a column of 10 cm, was equal to that reported by Almeida et al. [23, 24], higher than that reported elsewhere [6, 7, 61, 64], and lower than that reported by Kumar and Konwer [63]. In addition, the yields of gasoline, kerosene, light diesel, and heavy diesel of 1.55, 11.17, 21.38, and 32.72% (wt), obtained with a column of 10 cm, were in accordance with the yields of distillation fractions reported by Almeida et al. [23, 24] and Kumar and Konwer [63]. For the experiments carried out with OLP obtained with 5, 10, and 15% (wt) Na_2CO_3, using a column of 50 cm height, with and without reflux, the yields of distillates (biofuels) and gas increased in a sigmoid and linear fashion, respectively, with increasing catalyst content, while those of bottoms products decreased in a sigmoid fashion, as shown in **Figure 3**. Dandik and Aksoy [7] observed the same tendency. The yield of distillates obtained with 15% (wt) Na_2CO_3 and 50 cm column height (62.15%) was higher than that reported by Dandik and

Process parameters	Distillation without reflux			Distillation without reflux		
	15% (wt) Na_2CO_3			Column height 50 cm		
	Column height (cm)			(wt%) Na_2CO_3		
	10	30	50	5	10	15
Initial temperature (°C)	26	26	26	26	26	26
Final temperature (°C)	400	400	400	400	400	400
Processing time (min)	85	91	104	102	112	104
Distillation fractions $T^{B,I}$ (°C)						
(40ºC < T^B < 175ºC)	158	166	165	–	174	165
(175ºC < T^B < 235ºC)	190	185	190	180	188	190
(235ºC < T^B < 305ºC)	241	241	244	251	244	244
(305ºC < T^B < 400ºC)	310	311	311	308	315	311
Distillation fractions (material balances)						
Mass of feed (g)	694.10	636.05	685.21	692.63	660.69	685.21
Mass distillation fraction (40°C < T^B < 175ºC) (g)	10.75	14.24	12.03	–	4.77	12.03
Mass of aqueous phase (g)	0	0.17	0	–	0	0
Mass distillation fraction (175°C < T^B < 235ºC) (g)	77.55	78.74	79.13	20.71	57.50	79.13
Mass of aqueous phase (g)	0	0	0	4.08	0.92	0
Mass distillation fraction (235°C < T^B < 305ºC) (g)	148.41	136.32	138.60	108.30	106.04	138.60
Mass of aqueous phase (g)	0	0	0	0	0	0
Mass distillation fraction (305°C < T^B < 400ºC) (g)	227.14	186.73	196.14	115.05	203.33	196.14
Mass bottoms products (raffinate) (g)	205.48	197.74	235.84	426.24	269.02	235.84

http://dx.doi.org/10.5772/66759

Process parameters	Distillation without reflux			Distillation without reflux		
	15% (wt) Na_2CO_3			Column height 50 cm		
	Column height (cm)			(wt%) Na_2CO_3		
	10	30	50	5	10	15
Mass of gas (g)	24.77	22.11	23.47	18.25	18.81	23.47
Yield of gasoline-like fraction (wt%)	1.55	2.26	1.75	–	0.72	1.75
Yield of kerosene-like fraction (wt%)	11.17	12.38	11.55	3.58	8.85	11.55
Yield of light diesel-like fraction (wt%)	21.38	21.43	20.23	15.63	16.05	20.23
Yield of heavy diesel-like fraction (wt%)	32.72	29.35	28.62	16.61	30.79	28.62
Yield of biofuels (wt%)	66.83	65.43	62.15	35.82	56.41	62.15
Yield of gas (wt%)	3.57	3.47	3.43	2.63	2.85	3.43
Yield of raffinate (wt%)	29.60	31.10	34.42	61.54	40.74	34.42

$T^{B,I}$, initial boiling temperature; T^{B}, boiling temperature.

Table 3. Mass balances and yields of distillation products obtained by laboratory fractional distillation of OLP produced at 450°C, 1.0 atm, with 5, 10, and 15% (wt) Na_2CO_3, using Vigreux columns of 10, 30, and 50 cm, operating without reflux.

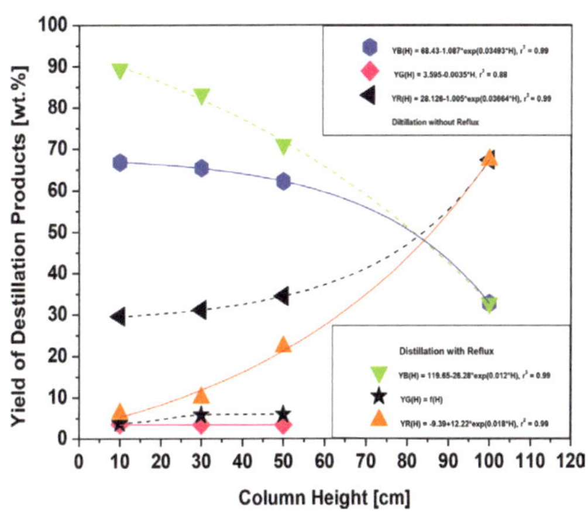

Figure 2. Yield of distillation products (distillates, bottoms, and gas), produced by laboratory distillation with and without reflux (YB, YR, and YG) of OLP obtained at 450°C and 1.0 atm, with 15% (wt) Na_2CO_3 in pilot scale, using columns of 10, 30, and 50 cm, and a pilot-packed distillation column of 100 cm height.

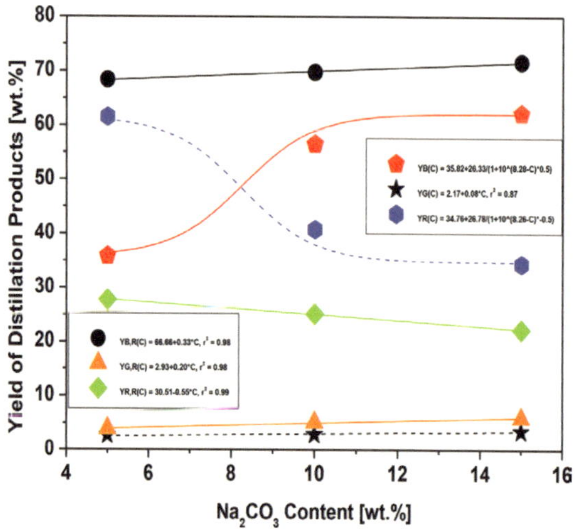

Figure 3. Yield of distillation products (distillates, bottoms, and gas), produced by laboratory distillation with (YB,R, YR,R, and YG,R) and without reflux (YB, YR, and YG) with OLP obtained at 450°C and 1.0 atm, with 5, 10, and 15% (wt) Na_2CO_3 in pilot scale, using a column of 50 cm.

Aksoy [7] at 420°C, 1.0 atm, with 10% (wt) Na_2CO_3, using a fractionating column of 54 cm, but lower than the one obtained by Kumar and Konwer [63] using an Oldershaw column of 50 cm.

3.3.1.2. Distillation with reflux: material balances and yields of distillation fractions

Table 4 shows the material balances and yields of distillation products (distillates, bottoms, and gas) produced by laboratory fractional distillation of OLP obtained at 450°C and 1.0 atm, with 5, 10, and 15% (wt) Na_2CO_3 in pilot scale, using Vigreux columns of different heights ($L_1 = 10$ cm, $L_2 = 30$ cm, $L_3 = 50$ cm), operating with reflux. The results show higher distillate yields and lower bottoms products yields compared to the fractional distillation without reflux, as well as the absence of heavy diesel-like fractions. In addition, the same tendency was observed for the variation of distillates, bottoms products, and gas yields with increasing column heights by fractional distillation of OLP obtained with 15% (wt) Na_2CO_3 and with a 50 cm column by fractional distillation of OLP obtained with 5, 10, and 15% (wt) Na_2CO_3. For the experiments with different column heights, a maximum distillate yield of 89.44% (wt) was achieved at 10 cm, much higher than those reported elsewhere [6, 7, 23, 24, 61, 63, 64], showing that reflux has improved the yields of distillates. This is according to the results of Kumar and Konwer [63] for the global yield of distillation fractions collected between 180 and 300°C, 300 and 325°C, and 325 and 370°C, operating with a reflux ratio of 0.2 and 10 mm Hg, obtaining 56.80% (wt). In addition, the yields of gasoline, kerosene, and light diesel of 10.86, 15.38, and 63.18% (wt) were according to the yields of gasoline (14.32%), kerosene (8.67%), and diesel (56.80%) reported by Kumar and Konwer [63]. For the experiments using a column of 50 cm height and OLP obtained with 5, 10, and 15% (wt) Na_2CO_3, a maximum distillate yield of 71.65% (wt) was achieved for OLP obtained with 15% (wt) Na_2CO_3, much higher than those reported elsewhere [6, 7, 23, 24, 61, 64], but lower than those reported by Kumar and Konwer [63].

Process parameters	Distillation with reflux			Distillation with reflux		
	15% (wt) Na_2CO_3			Column height 50 cm		
	Column height (cm)			(wt%) Na_2CO_3		
	10	**30**	**50**	**5**	**10**	**15**
Initial temperature (°C)	25	26	25	27	25	25
Initial reflux temperature (°C)	51	53	54	66	60	54
Initial reflux time (min)	10	10	11	18	15	11
Final temperature (°C)	305	305	305	270	305	305
Processing time (min)	270	243	250	257	235	250
Distillation fractions $T^{B,I}$ (°C)						
(40ºC < T^B < 175ºC)	118	113	90	104	84	90
(175ºC < T^B < 235ºC)	181	187	186	194	186	186
(235ºC < T^B < 305ºC)	239	240	244	245	236	244
(305ºC < T^B < 400ºC)	–	–	–	–	–	–
Distillation fractions (material balances)						
Mass of feed (g)	300	200	200	200	200	200
Mass distillation fraction (40°C < T^B < 175ºC) (g)	32.60	27.33	20.69	10.30	24.78	20.69
Mass of aqueous phase (g)	0	0.40	0.30	2.60	1.33	0.30
Mass distillation fraction (175°C < T^B < 235ºC) (g)	46.16	24.11	24.98	16.35	20.17	24.98
Mass of aqueous phase (g)	0	0	0	1.31	0.15	0
Mass distillation fraction (235°C < T^B < 305ºC) (g)	189.56	115.17	97.64	105.00	92.09	97.64
Mass of aqueous phase (g)	1.90	1.37	0	1.20	1.00	0
Mass distillation fraction (305°C < T^B < 400ºC) (g)	–	–	–	–	–	–
Mass bottoms products (raffinate) (g)	18.56	20.00	44.49	55.44	50.27	44.49

Process parameters	Distillation with reflux			Distillation with reflux		
	15% (wt) Na_2CO_3			Column height 50 cm		
	Column height (cm)			(wt%) Na_2CO_3		
	10	30	50	5	10	15
Mass of gas (g)	11.22	11.62	11.90	7.80	10.21	11.90
Yield of gasoline-like fraction (wt%)	10.86	13.66	10.34	6.45	13.05	10.34
Yield of kerosene-like fraction (wt%)	15.38	12.05	12.49	8.83	10.16	12.49
Yield of light diesel-like fraction (wt%)	63.18	57.58	48.82	53.10	46.51	48.82
Yield of heavy diesel-like fraction (wt%)	–	–	–	–	–	–
Yield of biofuels (wt%)	89.44	83.30	71.65	68.38	69.77	71.65
Yield of gas (wt%)	3.74	5.81	5.95	3.90	5.10	5.95
Yield of raffinate (wt%)	6.18	10.00	22.24	27.72	25.13	22.24

$T^{B,I}$, initial boiling temperature; T^{B}, boiling temperature.

Table 4. Mass balances and yields of distillation products produced by laboratory fractional distillation of OLP obtained at 450°C and 1.0 atm, with 5, 10, and 15% (wt) Na_2CO_3 in pilot scale, using Vigreux columns of 10, 30, and 50 cm, operating with reflux.

3.3.2. Pilot unit

Material balances and yields of distillation products produced by pilot fractional distillation of OLP, obtained at 450°C and 1.0 atm, with 15% (wt) Na_2CO_3 in pilot scale, using a differential distillation apparatus, packed with borosilicate-glass raschig rings of cylindrical geometry (ID = 1.0 cm, L = 1.0 cm), of 100 cm height, with internal reflux, are illustrated in **Table 5**. The results show a distillates yield of 32.68% (wt), higher than that reported by Dandik and Aksoy [7] at 400 and 420°C, column height of 54 cm, with 1, 5, and 10% (wt) Na_2CO_3, but lower than the one obtained by Kumar and Konwer [63], collected between 40 and 140°C, 140 and 180°C, and 180 and 300°C, being the last fraction performed with a reflux ratio of 0.2 and 10 mm Hg. The yield of distillates in pilot distillation scale was lower because of the higher column height, and the fact that distillation was carried out up to 280°C because of equipment instabilities. The distillation of OLP, obtained at 450°C and 1.0 atm, with 15% (wt) Na_2CO_3, using a differential distillation apparatus, packed with borosilicate-glass raschig rings, of 100 cm height, with internal reflux, improved the quality (physicochemical properties) of gasoline, kerosene, and light diesel-like hydrocarbon fractions, particularly the acid values. The acid values ranged between 0.334 and 0.420 mg KOH/g, below the maximum permitted (0.5 mg KOH/g) acid value limit specification for diesel fuel S10 of ANP 65 [67].

Process parameters	Column height (cm)
	100
Initial temperature (°C)	30
Final temperature (°C)	305
Processing time (min)	270
Distillation fractions $T^{B,I}$ (°C)	
(40ºC < T^B < 175ºC)	94.6
(175ºC < T^B < 235ºC)	174.9
(235ºC < T^B < 280ºC)	233.8
Distillation fractions (material balances)	
Mass of feed (g)	6100.00
Mass distillation fraction (40°C < T^B < 175ºC) (g)	241.34
Mass distillation fraction (175°C < T^B < 235ºC) (g)	631.24
Mass distillation fraction (235°C < T^B < 280ºC) (g)	1121.15
Mass bottoms products (raffinate) (g)	4106.27
Yield of gasoline-like fraction (wt%)	3.95
Yield of kerosene-like fraction (wt%)	10.35
Yield of light diesel-like fraction (wt%)	18.38
Yield of heavy diesel-like fraction (wt%)	–
Yield of biofuels (wt%)	32.68
Yield of raffinate (wt%)	67.32

$T^{B,I}$, initial boiling temperature; T^B, boiling temperature.

Table 5. Mass balances and yields of distillation products (distillates and bottoms) produced by pilot fractional distillation of OLP obtained at 450°C and 1.0 atm, with 15% (wt) Na_2CO_3 in pilot scale, using a differential distillation-packed column of 100 cm, with internal reflux.

3.4. Physicochemical properties of distillation fractions

3.4.1. Density of distillation fractions

Physicochemical properties of hydrocarbon-like fractions, produced by laboratory fractional distillation of OLP, using Vigreux columns of different heights (L_1 = 10 cm, L_2 = 30 cm, L_3 = 50 cm), operating with and without reflux, and a pilot differential distillation column, packed with borosilicate-glass raschig rings, of 100 cm height, with internal reflux, are illustrated in **Tables 6–8**. The density of distillation fractions, produced by laboratory distillation of OLP at 450°C and 1.0 atm, with 15% (wt) Na_2CO_3, with and without reflux using columns of different heights (L_1 = 10 cm, L_2 = 30 cm, L_3 = 50 cm), and a pilot-packed distillation column of 100 cm, with internal reflux, is shown in **Figure 4**. One may observe that densities of distillation fractions increase with increasing boiling temperature intervals, as reported by Kumar and Konver [63], remaining

Physicochemical properties	Distillation without reflux			Distillation without reflux		
	15% (wt) Na_2CO_3			Column height 50 cm		
	Column height (cm)			(wt%) Na_2CO_3		
	10	30	50	5	10	15
Distillation fraction (40ºC < T^B < 175ºC)						
ρ (g/cm³)	0.7312	0.7389	0.7376	–	0.7311	0.7376
I.A (mg KOH/g)	5.29	3.90	1.99	–	4.90	1.99
I.S (mg KOH/g)	8.76	12.86	35.81	–	–	35.81
I.R (–)	1.400	1.413	1.415	–	1.418	1.415
C (1A)	1A	1A	1A	–	1A	1A
Distillation fraction (175ºC < T^B < 235ºC)						
ρ (g/cm³)	0.7536	0.7497	0.7492	0.7298	0.7418	0.7492
μ (cSt)	0.88	0.84	0.81	0.72	0.76	0.81
I.A (mg KOH/g)	1.96	1.65	1.49	78.46	2.96	1.49
I.S (mg KOH/g)	10.56	6.39	11.87	95.17	5.27	11.87
I.R (–)	1.412	1.412	1.419	1.418	1.425	1.419
C (1A)	1A	1A	1A	1A	1A	1A
Distillation fraction (235ºC < T^B < 305ºC)						
ρ (g/cm³)	0.7887	0.7904	0.7873	0.7945	0.7833	0.7873
μ (cSt)	1.81	1.61	1.59	1.62	1.37	1.59
I.A (mg KOH/g)	1.47	1.07	0.98	49.09	2.62	0.98
I.S (mg KOH/g)	27.51	3.78	13.63	51.74	9.13	13.63
I.R (–)	1.442	1.441	1.442	1.441	1.439	1.442

http://dx.doi.org/10.5772/66759

Physicochemical properties	Distillation without reflux			Distillation without reflux		
	15% (wt) Na_2CO_3			Column height 50 cm		
	Column height (cm)			(wt%) Na_2CO_3		
	10	**30**	**50**	**5**	**10**	**15**
C (1A)	1A	1A	1A	1A	1A	1A
Distillation fraction (305°C < T^B <400°C)						
ρ (g/cm³)	0.8298	0.8470	0.8267	0.8189	0.8246	0.8267
μ (cSt)	4.98	4.40	4.03	5.07	4.34	4.03
I.A (mg KOH/g)	3.57	3.37	2.98	45.28	3.42	2.98
I.S (mg KOH/g)	30.42	64.62	40.23	64.10	44.61	40.23
I.R (–)	1.450	1.448	1.500	1.445	1.445	1.500
C (1A)	1A	1A	1A	1A	1A	1A

I.A, acid value; I.R, refractive index; I.S, saponification value; C, copper corrosiveness.

Table 6. Physicochemical analysis of hydrocarbon-like fractions produced by laboratory fractional distillation of OLP obtained at 450°C and 1.0 atm, with 5, 10, and 15% (wt) Na_2CO_3 in pilot scale, using Vigreux columns of 10, 30, and 50 cm, operating without reflux.

Physicochemical properties	Distillation with reflux 15% (wt) Na_2CO_3 Column height (cm)			Distillation with reflux Column height 50 cm (wt%) Na_2CO_3		
	10	30	50	5	10	15
Distillation fraction (40ºC < T^B < 175ºC)						
ϱ (g/cm³)	0.7494	0.7497	0.7492	0.8030	0.7530	0.7492
μ (cSt)	1.373	1.259	1.251	–	–	–
I.A (mg KOH/g)	2.08	2.57	2.96	94.42	3.14	2.96
I.S (mg KOH/g)	5.29	5.29	5.73	36.34	22.45	5.73
I.R (–)	1.421	1.423	1.425	1.449	1.429	1.425
C (1A)	1A	1A	1A	1A	1A	1A
Distillation fraction (175ºC < T^B < 235ºC)						
ϱ (g/cm³)	0.782	0.787	0.788	0.8090	0.7820	0.788
μ (cSt)	1.845	1.778	1.733	–	–	–
I.A (mg KOH/g)	2.94	2.75	1.65	80.2	3.40	1.65
I.S (mg KOH/g)	9.27	5.28	5.22	48.89	19.82	5.22
I.R (–)	1.439	1.440	1.442	1.445	1.441	1.442
C (1A)	1A	1A	1A	1A	1A	1A
Distillation fraction (235ºC < T^B < 305ºC)						
ϱ (g/cm³)	0.823	0.824	0.827	0.8190	0.8090	0.827
μ (cSt)	3.610	3.528	3.245	–	–	–
I.A (mg KOH/g)	4.68	3.90	3.39	49.09	5.33	3.39
I.S (mg KOH/g)	38.20	43.60	48.63	39.20	25.18	48.63
I.R (–)	1.451	1.451	1.453	1.454	1.451	1.453
C (1A)	1A	1A	1A	1A	1A	1A

I.A, acid value; I.R, refractive index; I.S, saponification value; C, copper corrosiveness.

Table 7. Physicochemical analysis of distillation fractions produced by laboratory fractional distillation of OLP obtained at 450°C and 1.0 atm, with 5, 10, and 15% (wt) Na_2CO_3 in pilot scale, using Vigreux columns of 10, 30, and 50 cm, operating with reflux.

http://dx.doi.org/10.5772/66759

Physicochemical properties	Pilot column
	Column height (cm)
	100
Distillation fraction (40ºC < T^B < 175ºC)	
ρ (g/cm³)	0.7171
μ (cSt)	0.59
I.A (mg KOH/g)	0.334
I.S (mg KOH/g)	10.52
I.R (–)	1.401
C (1A)	1A
Distillation fraction (175ºC < T^B < 235ºC)	
ρ (g/cm³)	0.7512
μ (cSt)	0.85
I.A (mg KOH/g)	0.42
I.S (mg KOH/g)	9.25
I.R (–)	1.420
C (1A)	1A
Distillation fraction (235ºC < T^B < 280ºC)	
ρ (g/cm³)	0.7862
μ (cSt)	1.52
I.A (mg KOH/g)	0.34
I.S (mg KOH/g)	10.56
I.R (–)	1.439
C (1A)	1

I.A, acid value; I.R, refractive index; I.S, saponification value; C, copper corrosiveness.

Table 8. Physicochemical analysis of distillation fractions produced by pilot distillation with internal reflux, of OLP obtained at 450°C and 1.0 atm, with 15% (wt) Na_2CO_3 in pilot scale, using a differential distillation-packed column of 100 cm height.

almost constant along with the column height for the experiments carried out in laboratory scale, with and without reflux. For the distillation experiments carried out in laboratory scale without reflux, a total of four hydrocarbon-like fractions were collected (gasoline, kerosene, light diesel, and heavy diesel), while for the experiments under reflux conditions, only three hydrocarbon-like fractions could be collected (gasoline, kerosene, and light diesel). This is probably because of the recycling of part of the distillates back into the distillation column. In addition, the densities of gasoline, kerosene, and light diesel produced by fractional distillation in laboratory scale with reflux superposed exactly those of kerosene, light diesel, and heavy diesel produced by fractional distillation in laboratory scale without reflux, showing the importance of operating under reflux

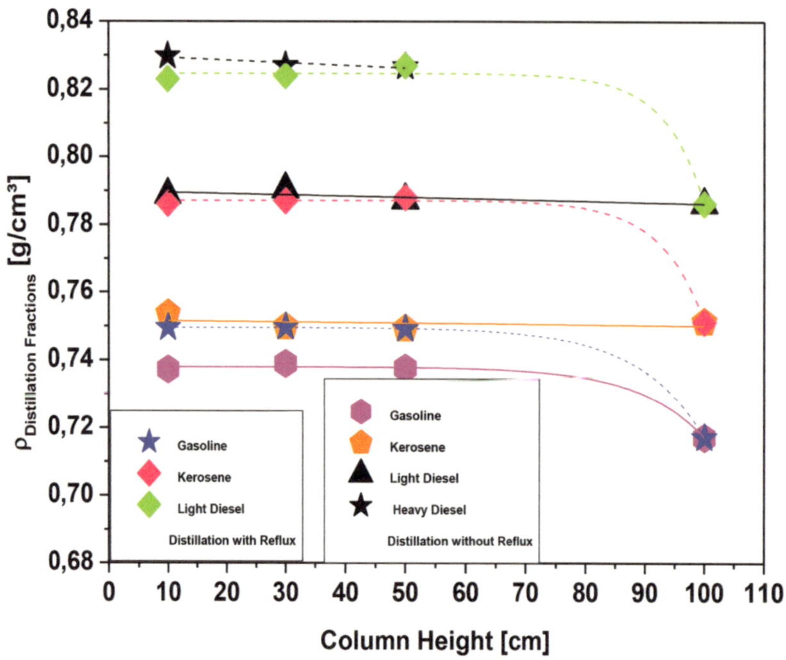

Figure 4. Density of hydrocarbon-like fractions produced by laboratory distillation of OLP obtained at 450°C and 1.0 atm, with 15% (wt) Na_2CO_3, with and without reflux using columns of 10, 30, and 50 cm, and a pilot-packed distillation column of 100 cm.

conditions to separate properly the hydrocarbon-like fractions. The densities of hydrocarbon-like fractions produced by fractional distillation in pilot scale, using a differential distillation column, packed with borosilicate-glass raschig rings, of 100 cm height, were lower in comparison to those produced by fractional distillation in laboratory scale, with and without reflux. Finally, the use of reflux made it possible to cut the hydrocarbon-like fractions properly, correcting the lower density limits, as observed by Almeida et al. [22–24], and thus matching the densities of kerosene and diesel fuels according to kerosene aviation specifications (QVA-1/JET A-1) of ANP 37 [68] and diesel S10 specification of ANP 65 [67].

3.4.2. Acid values of distillation fractions

The acid values of hydrocarbon-like fractions, produced by laboratory distillation of OLP (450°C and 1.0 atm, with 15% (wt) Na_2CO_3), without reflux using columns of different heights (L_1 = 10 cm, L_2 = 30 cm, L_3 = 50 cm), and a pilot-packed distillation column of 100 cm, with internal reflux, are illustrated in **Figure 5**. The acid values of hydrocarbon-like fractions decreased in a linear fashion with increasing column height for the experiments carried out in laboratory scale, without reflux, as shown in **Figure 5**. This is probably caused by the concentration of lighter volatile compounds in the vapor phase with increasing column height, so that the chemical compounds conferring the acidity of hydrocarbon-like fractions, particularly those of medium and long carbon chain length present in OLP, cannot reach the top of the distillation column, being present in small concentrations in the gaseous phase. The

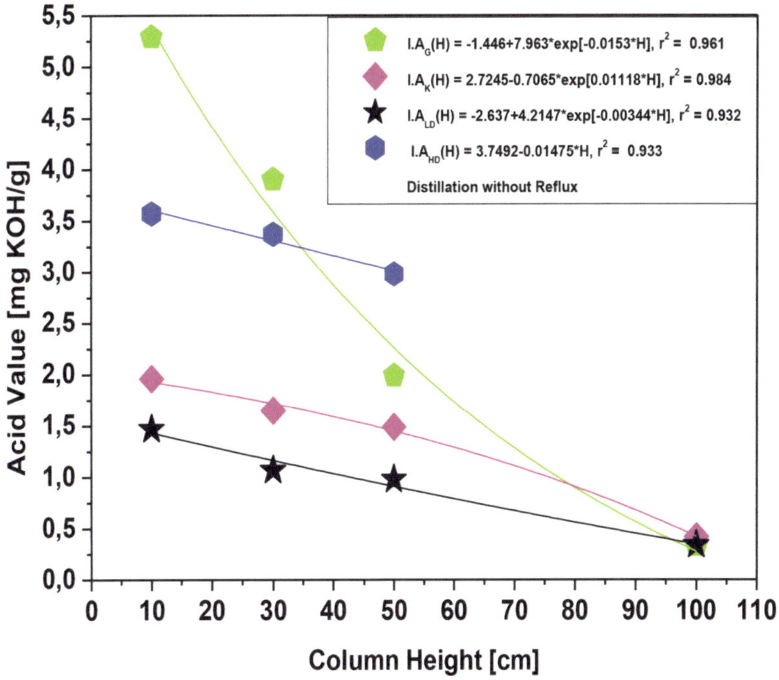

Figure 5. Acid values of distillation fractions produced by laboratory distillation of OLP obtained at 450°C and 1.0 atm, with 15% (wt) Na_2CO_3, without reflux using columns of 10, 30, and 50 cm, and a pilot-packed distillation column of 100 cm height.

acid values of distillation fractions also decreased with increasing boiling temperature ranges, except the heavy diesel-like fraction, which is in accordance with the results reported by Elkasabi et al. [64], for acid values of tail-gas reactive pyrolysis (TGRP) distillation fractions. The acid values of hydrocarbon-like fractions decreased with increasing Na_2CO_3 content, for distillation experiments in laboratory scale, using a column of 50 cm, with and without reflux, showing that fractional distillation of OLP with high acid values was ineffective. The acid values of hydrocarbon-like fractions produced by fractional distillation in pilot scale, using a differential distillation column, packed with borosilicate-glass raschig rings, of 100 cm height, were lower in comparison to those produced by fractional distillation in laboratory scale, with and without reflux. This showed that use of packed distillation columns improved not only the de-acidification process, but also the physicochemical properties of distillation fractions.

3.5. Chemical analysis of OLP and distillation fractions

3.5.1. FT-IR of OLP and distillation fractions

Figures 6–8 illustrate the FT-IR analysis of OLP obtained at 450°C and 1.0 atm, with 5, 10, and 15% (wt) Na_2CO_3, hydrocarbon-like fractions, produced by fractional distillation, using a pilot-packed distillation column of 100 cm height, and light diesel-like fractions, produced by laboratory distillation, using columns of 10, 30, and 50 cm height, without reflux. The

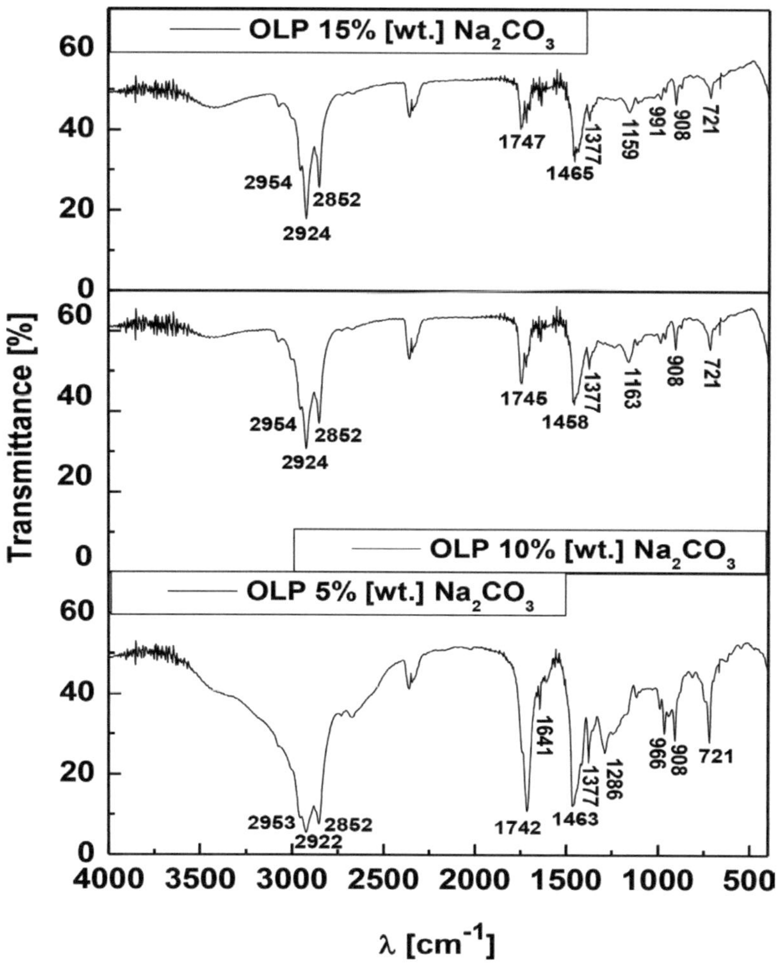

Figure 6. FT-IR of OLP obtained by catalytic cracking of palm oil at 450°C and 1.0 atm, with 5, 10, and 15% (wt) Na_2CO_3 in pilot scale.

identification of absorption bands/peaks was done according to previous studies [21, 24]. The spectrum of OLP obtained with 5% (wt) Na_2CO_3 presented a wide band of axial deformation at 3435 cm^{-1} compared to OLP obtained with 10 and 15% (wt) Na_2CO_3, characteristic of O–H intramolecular hydrogen bond, indicating probably the presence of fatty alcohols and/or carboxylic acids. This band was also observed for gasoline and kerosene-like fractions, using a pilot-packed distillation column of 100 cm height, as well as light diesel-like fraction, using columns of 10, 30, and 50 cm height, without reflux, both obtained by distillation of OLP at 450°C and 1.0 atm, with 15% (wt) Na_2CO_3. The spectra of OLP and distillation fractions exhibited intense peaks between 2921 and 2964 cm^{-1} and between 2858 and 2964 cm^{-1}, indicating the presence of aliphatic compounds associated with methylene (CH_2) and methyl (CH_3) groups. This confirmed the presence of hydrocarbons [21–24]. One can also observe for OLP and distillation fractions, except for light diesel-like fraction, produced by laboratory distillation without reflux, using columns of 10 cm, the presence of bands at

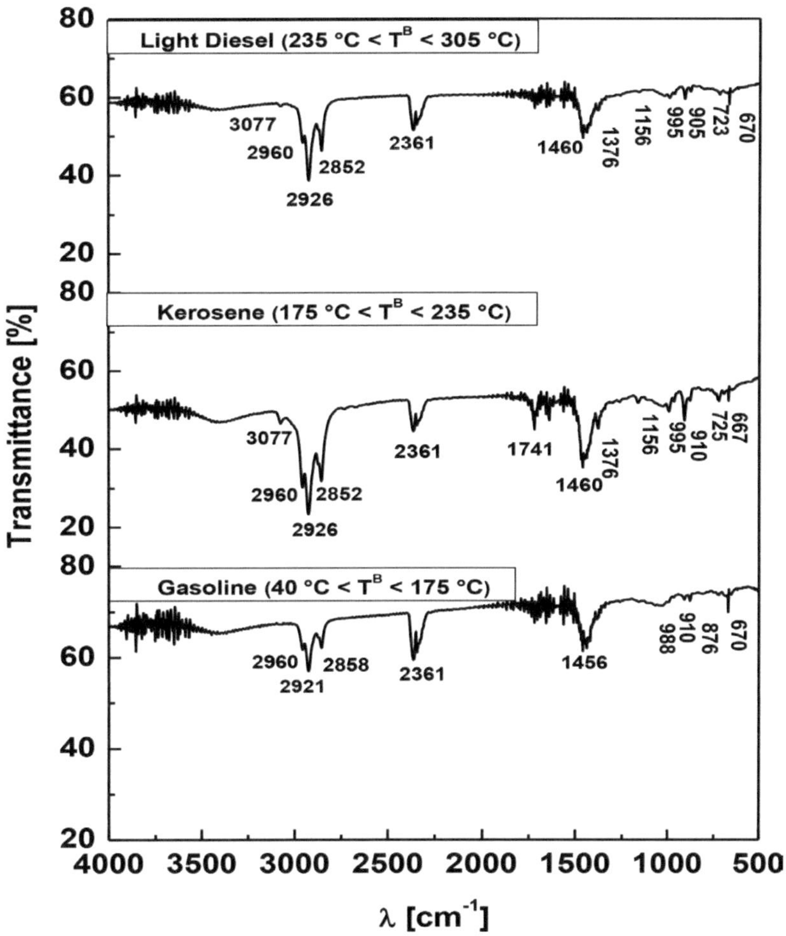

Figure 7. FT-IR of hydrocarbon-like fractions produced in a pilot-packed distillation column with internal reflux of 100 cm height with OLP obtained at 450°C and 1.0 atm, with 15% (wt) Na_2CO_3 in pilot scale.

2361 cm^{-1}, characteristic of asymmetrical axial deformation of CO_2. In addition, both OLP obtained at 450°C and 1.0 atm, with 5% (wt) Na_2CO_3, exhibited the presence of an intense and larger axial deformation band between 3200 and 2500 cm^{-1}, characteristic of hydroxyl (O–H) groups [39, 40], indicating the absence of carboxylic acids. This is according to the OLP acid value of 51.56 mg KOH/g. An intense axial deformation band has been observed for OLP, whose intensity decreases with Na_2CO_3 content, characteristic of carbonyl (C=O) groups, with peaks at 1742, 1745, and 1747 cm^{-1} probably associated with a ketone and/or carboxylic acids [21–24]. This is according to the acid values of OLP presented in **Table 1**, with its highest value obtained with 5% (wt) Na_2CO_3. These peaks were also observed in kerosene, produced by pilot-scale distillation, and light diesel, produced by laboratory distillation without reflux, using columns of 30 cm. The spectra of OLP and distillation fractions were exhibited between 1455 and 1465 cm^{-1}, a characteristic asymmetrical deformation vibration of methylene (CH_2) and methyl (CH_3) groups, indicating the presence of alkanes [21–24]. The spectrum of OLP and distillation fractions was identified at 1377 cm^{-1}, except for light diesel, produced

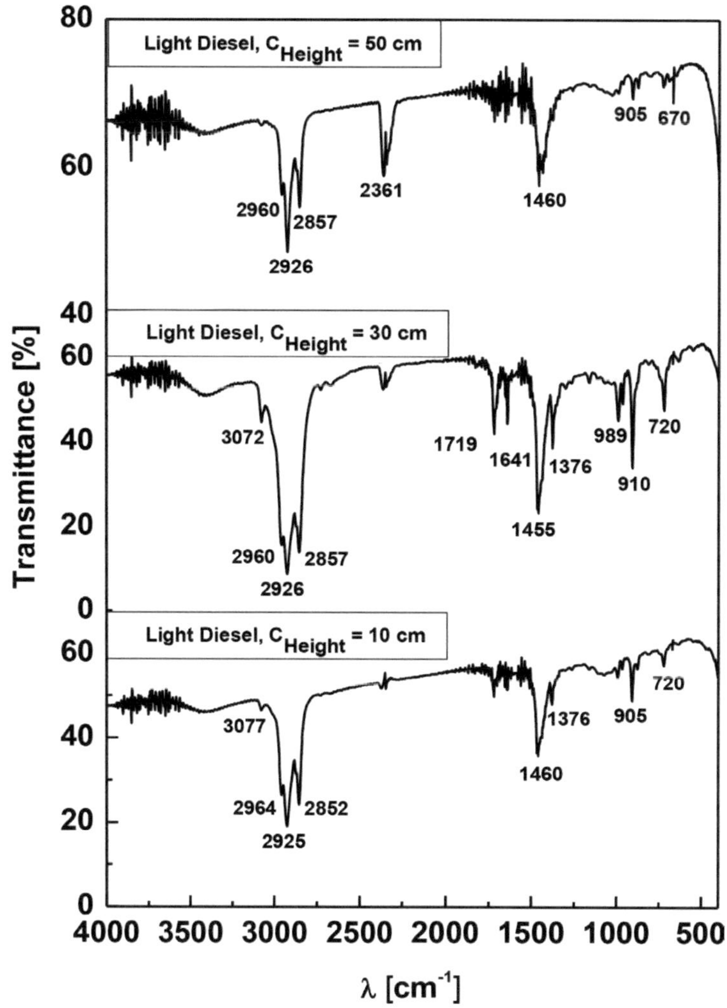

Figure 8. FT-IR of light diesel-like hydrocarbon fraction produced by laboratory distillation without reflux using columns of 10, 30, and 50 cm height with OLP obtained at 450°C and 1.0 atm, with 15% (wt) Na_2CO_3 in pilot scale.

by pilot-scale distillation and by laboratory distillation without reflux, using columns of 50 cm, a band of symmetrical angular deformation of C–H bonds in the methyl group (CH_3) [21–24]. The peaks between 995 and 905 cm^{-1} for OLP and distillation fractions were characteristic of an angular deformation outside the plane of C–H bonds, indicating the presence of alkenes [21–24]. The spectra of OLP and light diesel fraction exhibited bands between 721 and 667 cm^{-1}, peaks characteristic of an angular deformation outside the plane of C–H bonds in the methylene (CH_2) group, indicating the presence of olefins [21–24]. The FT-IR analysis of OLP identified the presence of aliphatic groups (alkenes, alkanes, etc.), as well as oxygenates (carboxylic acids, ketones, fatty alcohols), and the presence of aliphatic groups (alkenes, alkanes, etc.) in light diesel fraction.

3.5.2. GC-MS of OLP and light diesel-like hydrocarbon fraction

Figures 9 and **10** illustrate the chromatograms of OLP obtained by catalytic cracking of palm oil at 450°C and 1.0 atm, with 15% (wt) Na_2CO_3 in pilot scale and light diesel-like hydrocarbon fraction produced by fractional distillation, using a pilot-packed distillation column with internal reflux of 100 cm height. The classes of compounds, summation of peak areas, and retention times of chemical compounds identified by GC-MS of OLP obtained at 450°C and 1.0 atm, with 15% (wt) Na_2CO_3 and light diesel-like fraction produced by pilot fractional distillation of OLP, using a differential distillation column of 100 cm height, are described in **Table 9**. GC-MS detected the presence of hydrocarbons such as alkenes, alkanes, alkynes, ring-containing alkenes, ring-containing alkanes, and dienes, as well as oxygenates such as ketones and fatty alcohols. OLP is composed of 92.84% (area) hydrocarbons (52.72% alkenes, 27.53% alkanes, 4.20% ring-containing alkenes, 6.33% ring-containing alkanes, and 1.21% dienes) and 7.16% (area) oxygenates (4.50% ketones and 2.66% fatty alcohols), while the light diesel-like fraction is composed of 100% hydrocarbons (57.08% alkenes, 34.85% alkanes, and 8.07% ring-containing alkanes). In both OLP and light diesel-like fraction, GC-MS had not identified the presence of carboxylic acids. The results were in accordance with the low acid values of OLP (3.55 mg KOH/g) and light diesel-like fraction (0.34 mg KOH/g) presented in **Tables 1** and **8**. The concentration of hydrocarbons in OLP, expressed in % (area), was higher compared to similar studies reported in the literature [21–24]. The chemical composition of OLP indicated the presence of heavy gasoline compounds with C_9 and C_{10} (C_5–C_{10}), kerosene-like fractions (C_{11}–C_{12}), light diesel-like fractions (C_{13}–C_{17}), and light heavy diesel compounds with C_{18} and C_{19} (C_{18}–C_{25}), as reported in the literature [22–24]. The light diesel-like fraction presented carbon chain lengths between C_{10} and C_{20} with the following carbon chain lengths: alkenes C_{10}–C_{20}, alkanes C_{10}–C_{16}, and ring-containing alkanes C_{11}–C_{12}. It may be observed that

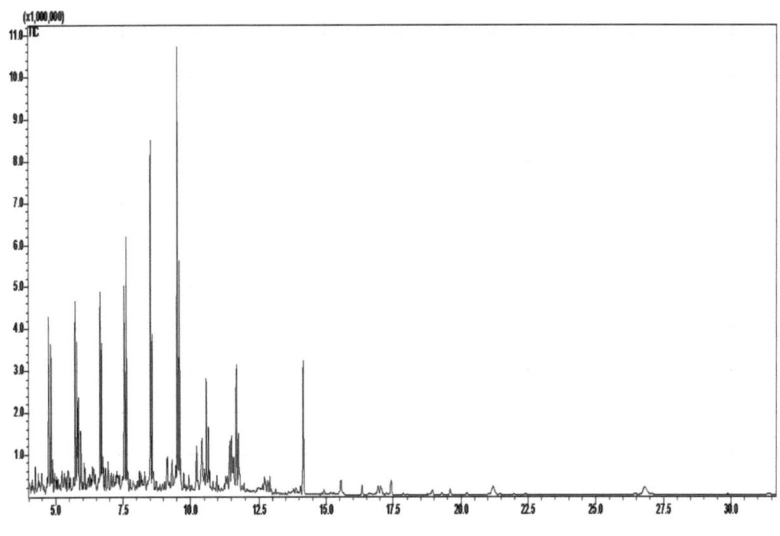

Figure 9. GC-MS of OLP obtained by catalytic cracking of palm oil at 450°C and 1.0 atm, with 15% (wt) Na_2CO_3 in pilot scale.

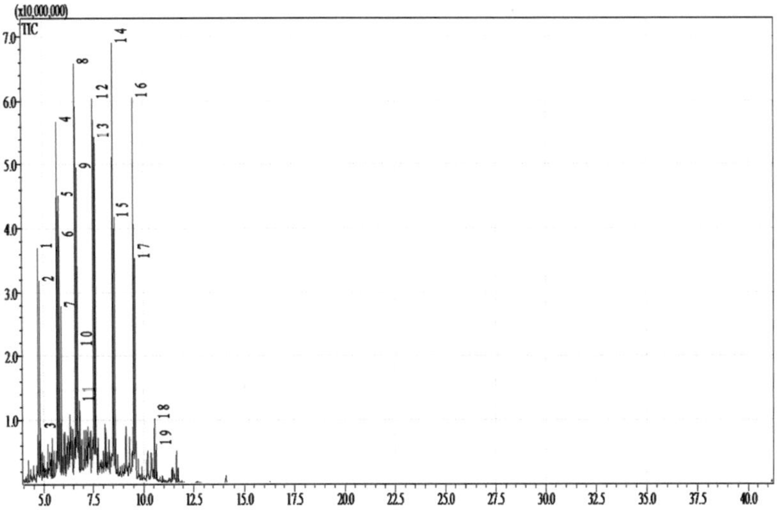

Figure 10. GC-MS of light diesel-like hydrocarbon fraction produced in a pilot-packed distillation column with internal reflux of 100 cm height with OLP obtained at 450°C and 1.0 atm, with 15% (wt) Na_2CO_3 in pilot scale.

OLP, 15% (wt)		**Light diesel-like fraction, 15% (wt)**	
Class of compounds: chemical compounds	**RT (min)**	**Class of compounds: chemical compounds**	**RT (min)**
Alkenes		*Alkenes*	
1-Decene	4.74	1-Decene	4.74
4-Decene	4.87	2-Decene	4.87
(3E)-3-Dodecene	5.92	1-Undecene	5.71
1-Dodecene	6.65	1-Dodecene	6.65
(2E)-2-Dodecene	6.76	5-Tetradecene	6.76
1-Tridecene	7.56	1-Tetradecene	7.56
1-Tetradecene	8.51	1-Pentadecene	8.51
(9E)-9-Octadecene	9.46	1-Heptadecene	9.52
1-Pentadecene	9.52	9-Eicosene	10.56
(3Z)-3-Hexadecene	10.49	*Σ (Area%) =*	57.08
1-Nonadecene	11.58	*Alkanes*	
1-Heptadecene	11.68	Decane	4.82
Z-5-Nonadecene	12.71	Undecane	5.79
Σ (Area%) =	**52.72**	Dodecane	6.72
Ring-containing alkenes		Tridecane	7.64
1-Propyl-1-cyclohexene	4.48	Tetradecane	8.58
1-Butylcyclohexene	5.45	Pentadecane	9.59

http://dx.doi.org/10.5772/66759

OLP, 15% (wt)		Light diesel-like fraction, 15% (wt)	
Class of compounds: chemical compounds	**RT (min)**	**Class of compounds: chemical compounds**	**RT (min)**
1-Octyl-1-cyclopentene	8.15	Hexadecane	10.63
1-Hexyl-1-cyclopentene	9.16	*Σ (Area%) =*	34.85
1-Decyl-1-cyclohexene	9.32	*Ring-containing alkanes*	
1-Hexyl-1-cyclohexene	10.41	1,2-Dibutyl-cyclopropane	5.83
Σ (Area%) =	***4.20***	1-Pentyl-2-propyl-cyclopropane	5.92
Alkanes		Nonyl-cyclopropane	6.84
Decane	4.82	*Σ (Area%) =*	8.07
Undecane	5.79		
Dodecane	6.72		
Tridecane	7.63		
Tetradecane	8.58		
Pentadecane	9.60		
Nonadecane	11.76		
Σ (Area%) =	***27.53***		
Ring-containing alkanes			
Isobutylcyclohexane	4.23		
1,2-Dimethylcyclooctane	4.96		
Butylcyclohexane	5.22		
1-Pentyl-2-propylcyclopropane	5.83		
Cyclododecane	6.85		
Decylcyclopentane	8.11		
Nonylcyclopentane	9.13		
Cyclopentadecane	9.64		
n-Nonylcyclohexane	10.22		
1-Pentyl-2-propylcyclopentane	11.80		
Σ (Area%) =	***6.33***		
Dienes			
4,6-Decadiene	5.08		
Z-1,6-Undecadiene	6.04		
(2E,4Z)-2,4-Dodecadiene	6.36		
Σ (Area%) =	***1.21***		

OLP, 15% (wt)		Light diesel-like fraction, 15% (wt)	
Class of compounds: chemical compounds	**RT (min)**	**Class of compounds: chemical compounds**	**RT (min)**
Alkynes			
6-Tridecyne	6.96		
Σ (Area%) =	***0.85***		
Ketones			
2-Pentadecanone	14.14		
2-Nonadecanone	17.42		
Σ (Area%) =	***4.50***		
Alcohols			
Oleyl alcohol	11.43		
Σ (Area%) =	***2.66***		

Table 9. Classes of compounds, summation of peak areas, and retention times of chemical compounds identified by CG-MS of OLP obtained at 450°C and 1.0 atm, with 15% (wt) Na_2CO_3 and light diesel-like fraction produced by pilot fractional distillation of OLP, using a differential distillation column of 100 cm height.

the presence of gasoline heavy compounds with C_{10} (C_5–C_{10}), kerosene fractions (C_{11}–C_{12}), and light heavy diesel compounds with C_{18}, C_{19}, and C_{20} (C_{18}–C_{25}) in light diesel-like fraction, showed that fractional distillation in a pilot-packed distillation column of 100 cm with internal reflux was not capable of proper separation of the hydrocarbon-like fractions. This is probably caused by the limitation of internal reflux.

4. Conclusions

The yields of distillates and gas decreased along with the column height, while that of bottoms products increased, for experiments with and without reflux. The yields of distillates and gas increased with increasing catalyst content, while that of bottoms products decreased. In addition, distillation under reflux conditions showed higher distillates yields and lower bottoms products yields compared to the fractional distillation without reflux, as well as the absence of heavy diesel-like fractions. The densities of distillation fractions increased with increasing boiling temperature intervals, remaining almost constant along with the column height. In addition, the densities of gasoline, kerosene, and light diesel produced by fractional distillation in laboratory scale with reflux superposed exactly those of kerosene, light diesel, and heavy diesel produced by fractional distillation in laboratory scale without reflux, showing the importance of operating under reflux. The use of reflux made it possible to cut the hydrocarbon-like fractions properly, correcting the lower density limits. The acid values of hydrocarbon-like fractions decreased with increasing column height for the experiments with and without reflux. The acid values of distillation fractions showed a tendency to decrease

http://dx.doi.org/10.5772/66759

with increasing boiling temperature ranges. In addition, acid values of distillation fractions decreased with increasing Na_2CO_3 content, for distillation experiments using a column of 50 cm, with and without reflux. The distillation experiments in pilot scale showed gasoline, kerosene, and light diesel-like acid values of 0.33, 0.42, and 0.34 mg KOH/g, proving that use of packed distillation columns improved not only the de-acidification process, but also the physicochemical properties of distillation fractions. FT-IR of OLP and distillation fractions identified the presence of aliphatic hydrocarbons (alkanes, alkenes, etc.) and the absence of carbonyl groups. The light diesel-like fraction was composed of 100% hydrocarbons with an acid value of 0.34 mg KOH/g, density of 0.7862 g/cm^3, and kinematic viscosity of 1.52 mm^2 s^{-1}, proving the technical feasibility of OLP de-acidification by the fractional distillation process.

Author details

C. C. Fereira[1,2], E. C. Costa[1,2], D. A. R. de Castro[1,2], M. S. Pereira[1,2,3], A. A. Mâncio[1,2], M. C. Santos[1,2], D. E. L. Lhamas[4], S. A. P. da Mota[5], M. E. Araújo[1], Luiz E. P. Borges[3] and N. T. Machado[1,2]*

*Address all correspondence to: machado@ufpa.br

1 Laboratory of Separation Processes and Applied Thermodynamic (TERM@), Faculty of Chemical Engineering-UFPA, Belém, Pará, Brazil

2 Graduate Program of Natural Resource Engineering-UFPA, Belém, Pará, Brazil

3 Laboratory of Catalyst Preparation and Catalytic Cracking. Section of Chemical Engineering-IME, Rio de Janeiro, RJ, Brazil

4 Faculty of Mining and Chemical Engineering-UNIFESSPA, Marabá, Pará, Brazil

5 Faculty of Materials Engineering-UNIFESSPA, Marabá, Pará, Brazil

References

[1] M. J. Haas. Animal Fats. In: F. Shahidi (Ed.). *Bailey's Industrial Oil and Fat Products.* 6th Edition, Hoboken, New Jersey: John Wiley & Sons, Inc., 2005, Vol. 5, 267–284

[2] F. D. Gunstone. *The Chemistry of Oils and Fats: Sources, Composition, Properties and Uses.* UK: Blackwell Publishing Limited, 2004

[3] K. D. Maher, D. C. Bressler. Pyrolysis of triglyceride materials for the production of renewable fuels and chemicals. *Bioresource Technology.* 98 (2007) 2351–2368

[4] E. Vonghia, D. G. B. Boocock, S. K. Konar, A. Leung. Pathways for the deoxygenation of triglycerides to aliphatic hydrocarbons over activated alumina. *Energy & Fuels.* 9 (1995) 1090–1096

[5] R. O. Idem, S. P. R. Katikaneni, N. N. Bakhshi. Thermal cracking of canola oil: reaction products in the presence and absence of steam, *Energy Fuels*. 10 (1996) 1150–1162

[6] L. Dandik, H. A. Aksoy, A. Erdem-Senatalar. Catalytic conversion of used oil to hydrocarbon fuels in a fractionating pyrolysis reactor. *Energy & Fuels*. 12 (1998) 1148–1152

[7] L. Dandik, H. A. Aksoy. Pyrolysis of used sunflower oil in the presence of sodium carbonate by using fractionating pyrolysis reactor. *Fuel Processing Technology*. 57 (2) (1998) 81–92

[8] D. G. Lima, V. C. D. Soares, E. B. Ribeiro, D. A. Carvalho, E. C. V. Cardoso, F. C. Rassi. Diesel-like fuel obtained by pyrolysis of vegetable oils, *Journal of Analytical and Applied Pyrolysis*. 71 (2004) 987–996

[9] T. Hua, L. Chunyi, Y. Chaohe, S. Honghong. Alternative processing technology for converting vegetable oil and animal fats to clean fuels and light olefins. *Chinese Journal of Chemical Engineering*. 16 (3) (2008) 394–400

[10] C. M. R. Prado, N. R. Antoniosi Filho. Production and characterization of the biofuels obtained by thermal cracking and thermal catalytic cracking of vegetable oils. *Journal of Analytical and Applied Pyrolysis*. 86 (2009) 338–347

[11] V. R. Wiggers, H. F. Meier, A. Wisniewski Jr., A. A. Chivanga Barros, M. R. Wolf Maciel. Biofuels from continuous fast pyrolysis of soybean oil: a pilot plant study. *Bioresource Technology*. 100 (2009) 6570–6577

[12] X. Junming, J. Jianchun, S. Yunjuan, C. Jie. Production of hydrocarbon fuels from pyrolysis of soybean oils using a basic catalyst. *Bioresource Technology*. 101 (2010) 9803–9806

[13] N. Taufiqurrahmi, S. Bhatia. Catalytic cracking of edible and non-edible oils for the production of biofuels. *Energy and Environmental Sciience* 4 (2011) 1087–1112

[14] E. Buzetzki, K. Sidorová, Z. Cvengrošová, J. Cvengroš. Effects of oil type on products obtained by cracking of oils and fats. *Fuel Processing Technology*. 92 (2011) 2041–2047

[15] E. Buzetzki, K. Sidorová, Z. Cvengrošová, A. Kaszonyi, J. Cvengroš. The influence of zeolite catalysts on the products of rapeseed oil cracking. *Fuel Processing Technology*. 92 (2011) 1623–1631

[16] S. Yan, C. Dimaggio, H. Wang, S. Mohan, M. Kim, L. Yang, S. O. Salley, K. Y. Simon Ng. Catalytic conversion of triglycerides to liquid biofuels through transesterification, cracking, and hydrotreatment processes. *Current Catalysis*. 1 (2012) 41–51

[17] F. Yu, L. Gao, W. Wang, G. Zhang, J. Ji. Bio-fuel production from the catalytic pyrolysis of soybean oil over Me-Al-MCM-41 (Me = La, Ni or Fe) mesoporous materials. *Journal of Analytical and Applied Pyrolysis*. 104 (2013) 325–329

[18] V. P. Doronin, O. V. Potapenko, P. V. Lipin, T. P. Sorokina. Catalytic cracking of vegetable oils and vacuum gas oil. *Fuel*. 106 (2013) 757–765

[19] V. R. Wiggers, G. R. Zonta, A. P. França, D. R. Scharf, E. L. Simionatto, L. Ender, H. F. Meier. Challenges associated with choosing operational conditions for triglyceride thermal cracking aiming to improve biofuel quality. *Fuel*. 107 (2013) 601–608

[20] L. Li, K. Quan, J. Xu, F. Liu, S. Liu, S. Yu, C. Xie, B. Zhang, X. Ge. Liquid hydrocarbon fuels from catalytic cracking of waste cooking oils using basic mesoporous molecular sieves K2O/Ba-MCM-41 as catalysts. *ACS Sustainable Chemistry and Engineering* 1 (11) (2013) 1412–1416

[21] S. A. P. Mota, A. A. Mancio, D. E. L. Lhamas, D. H. de Abreu, M. S. da Silva, W. G. dos Santos, D. A. R. de Castro, R. M. de Oliveira, M. E. Araújo, L. E. P. Borges, N. T. Machado. Production of green diesel by thermal catalytic cracking of crude palm oil (Elaeis guineensis Jacq) in a pilot plant. *Journal of Analytical and Applied Pyrolysis*. 110 (2014) 1–11

[22] H. da Silva Almeida, O. A. Correa, J. G. Eid, H. J. Ribeiro, D. A. R. de Castro, M. S. Pereira, L. M. Pereira, A. de Andrade Mancio, M. C. Santos, J. A. da Silva Souza, Luiz E. P. Borges, N. M. Mendonca, N. T. Machado. Production of biofuels by thermal catalytic cracking of scum from grease traps in pilot scale. *Journal of Analytical and Applied Pyrolysis*. 118 (2016) 20–33

[23] H. da Silva Almeida, O. A. Corrêa, J. G. Eid, H. J. Ribeiro, D. A. R. de Castro, M. S. Pereira, L. M. Pereira, A. de Andrade Aâncio, M. C. Santos, S. A. P da Mota, J. A. da Silva Souza, Luiz E. P. Borges, N. M. Mendonça, N. T. Machado. Performance of thermochemical conversion of fat, oils, and grease into kerosene-like hydrocarbons in different production scales. *Journal of Analytical and Applied Pyrolysis*. 120 (2016) 126–143. DOI: 10.1016/j.jaap.2016.04.017

[24] H. da Silva Almeida, O. A. Corrêa, C. C. Ferreira, H. J. Ribeiro, D. A. R. de Castro, M. S. Pereira, A. de Andrade Aâncio, M. C. Santos, S. A. P da Mota, J. A. da Silva Souza, Luiz E. P. Borges, N. M. Mendonça, N. T. Machado. Diesel-like hydrocarbon fuels by catalytic cracking of fats, oils, and grease (FOG) from grease traps. *Journal of the Energy Institute*, DOI: 10.1016/j.joei.2016.04.008.

[25] J. A. Melero, J. Iglesias, A. Garcia. Biomass as renewable feedstock in standard refinery units. Feasibility, opportunities and challenges. *Energy and Environmental Science*. 5 (2012) 7393–7420

[26] H. Lappi, R. Alén. Production of vegetable oil-based biofuels— thermochemical behavior of fatty acid sodium salts during pyrolysis. *Journal of Analytical and Applied Pyrolysis* 86 (2009) 274–280

[27] H. Lappi, R. Alén. Pyrolysis of vegetable oil soaps—palm, olive, rapeseed and castor oils. *Journal of Analytical and Applied Pyrolysis*. 91, (2011) 154–158

[28] S. Wang, Z. Guo, Q. Cai, L. Guo. Catalytic conversion of carboxylic acids in bio-oil for liquid hydrocarbons production. *Biomass Bioenergy*. 45 (2012) 138–143

[29] P. Bielansky, A. Weinert, C. Schönberger, A. Reichhold. Gasoline and gaseous hydrocarbons from fatty acids via catalytic cracking. *Biomass Conversion and Biorefinery*. 2 (2012) 53–61

[30] B. Weber, E. A. Stadlbauer, S. Stengl, M. Hossain, A. Frank, D. Steffens, E. Schlich, G. Schilling. Production of hydrocarbons from fatty acids and animal fat in the presence of water and sodium carbonate. Reactor performance and fuel properties. *Fuel*. 94 (2012) 262–269

[31] Y. S. Ooi, R. Zakaria, A. R. Mohamed, S. Bhatia. Catalytic conversion of palm oil-based fatty acid mixture to liquid fuel. *Biomass and Bioenergy*. 27 (5) (2004) 477–484

[32] Y. S. Ooi, R. Zakaria, A. R. Mohamed, S. Bhatia. Catalytic conversion of fatty acids mixture to liquid fuels over mesoporous materials. *Reaction Kinetics and Catalysis Letters*. 84 (2) (2005) 295–302

[33] Y. S. Ooi, R. Zakaria, A. R. Mohamed, S. Bhatia. Catalytic cracking of used palm oil and palm oil fatty acids mixture for the production of liquid fuel: kinetic modeling. *Journal of the American Chemical Society*. 18 (2004) 1555–1561

[34] Y.S. Ooi, R. Zakaria, A. R. Mohamed, S. Bhatia S. Catalytic conversion of fatty acids mixture to liquid fuel and chemicals over composite microporous/mesoporous catalysts. *Energy Fuels*. 19 (3) (2005) 736–743

[35] W. Charusiri, T. Vitidsant. Kinetic study of used vegetable oil to liquid fuels over sulfated zirconia. *Energy Fuels*. 19 (5) (2005) 1783–1789

[36] F. A. Twaiq, A. R. Mohamed, S. Bhatia. Liquid hydrocarbon fuels from palm oil by catalytic cracking over aluminosilicate mesoporous catalysts with various Si/Al ratios. *Microporous and Mesoporous Materials*. 64 (1–3) (2003) 95–107

[37] F. A. Twaiq, N. A. M. Zabidi, S. Bhatia. Catalytic conversion of palm oil to hydrocarbons: performance of various zeolite catalysts. *Industrial and Engineering Chemistry Research*. 38 (9) (1999) 3230–3237

[38] F. A. Twaiq, A. R. Mohamad, S. Bhatia. Performance of composite catalysts in palm oil cracking for the production of liquid fuels and chemicals. *Fuel Processing Technology*. 85 (11) (2004) 1283–1300

[39] D. Y. Siswanto, G. W. Salim, N. Wibisono, H. Hindarso, Y. Sudaryanto, S. Ismadji. Gasoline production from palm oil via catalytic cracking using MCM41: determination of optimum conditions. *Journal of Engineering and Applied Sciences*. 3 (6) (2008) 42–46

[40] P. K. Kanaujia, D. V. Naik, D. Tripathi, R. Singh, M. K. Poddar, L. N. Siva Kumar Konathala, Y. K. Sharma. Pyrolysis of *Jatropha Curcas* seed cake followed by optimization of liquid–liquid extraction procedure for the obtained bio-oil. *Journal of Analytical and Applied Pyrolysis*. (118) (2016) 202–224

[41] V. R. Wiggers, A. Wisniewski Jr., L. A. S. Madureira, A. A. Chivanga Barros, H. F. Meier. Biofuels from waste fish oil pyrolysis: continuous production in a pilot plant. *Fuel*. 88 (11) (2009) 2135–2141

[42] D. Konwer, S. E. Taylor, B. E. Gordon, J. W. Otvos, M. Calvin. Liquid fuels from *Mesua ferrea* L. seed oil. *JAOCS*. 66 (2) (1989) 223–226

[43] F. A. Zaher, A. R. Taman. Thermally decomposed cottonseed oil as a diesel-engine fuel. *Energy Sources*. 15 (3) (1993) 499–504

[44] A. de Andrade Mâncio. Production, fractionation and de-acidification of biofuels obtained by thermal catalytic cracking of vegetable oils. PhD Thesis, Graduate Program of Natural Resources Engineering, April 2015, UFPA, CDD 22, Ed. 660.2995, http://proderna.propesp.ufpa.br/ARQUIVOS/teses/Andreia.pdf

[45] A. L. F. Santos, D. U. Martins, O. K. Iha, R. A. M. Ribeiro, R. L. Quirino, P. A. Z. Suarez. Agro-industrial residues as low-price feedstock for diesel-like fuel production by thermal cracking. *Bioresource Technology*. 101 (15) (2010) 6157–6162

[46] T. Kraiem, A. B. Hassen-Trabelsi, S. Naoui, H. Belayouni, M. Jeguirim. Characterization of the liquid products obtained from Tunisian waste fish fats using the pyrolysis process. *Fuel Processing Technology*. 138 (2015) 404–412

[47] L. Dandik, H. A. Aksoy. Effect of catalyst on the pyrolysis of used oil carried out in a fractionating pyrolysis reactor. *Renewable Energy*. 16 (1–4) (1999) 1007–1010

[48] Y. S. Ooi. Biofuel production from catalytic cracking of palm oil. *Energy Sources*. 25 (9) (2003) 859–869

[49] Y. S. Ooi, R. Zakaria, A. R. Mohamed, S. Bhatia. Synthesis of composite material MCM-41/beta and its catalytic performance in waste used palm oil cracking. *Applied Catalysis A: General*. 274 (1–2) (2004) 15–23

[50] W. H. Chang, C. T. Tye. Catalytic cracking of used palm oil using composite zeolite. *Malaysian Journal of Analytical Sciences*. 17 (1) (2013) 176–184

[51] B. Weber, E. A. Stadlbauer, S. Eichenauer, A. Frank, D. Steffens., S. Elmar, S. Gerhard. Characterization of alkanes and olefins from thermo-chemical conversion of animal fat. *Journal of Biobased Materials and Bioenergy*. 8 (5) (2014) 526–537

[52] S. Eichenauer, B. Weber, E. A. Stadlbauer. Thermochemical processing of animal fat and meat and bone meal to hydrocarbons based fuels. ASME 2015, 9th International Conference on Energy Sustainability, Paper No. ES2015-49197, V001T02A001, DOI: 10.115/ES2015-49197

[53] S. A. Pereira da Mota. Thermal catalytic cracking of vegetable oils in different production scales. PhD Thesis, September 2013, UFPA, CDD 22, Ed. 660.2995, http://proderna.propesp.ufpa.br/ARQUIVOS/teses/Silvio.pdf

[54] D. E. Lima Lhamas. Study of biofuels production process by thermal catalytic cracking of palm oil (*Elaeis guineensis*) and buriti oil (*Mauritia flexuosa* L.). PhD Thesis, December 2013, UFPA, http://proderna.propesp.ufpa.br/ARQUIVOS/teses/Dyenny.pdf

[55] M. Costa Santos. Study of biofuels production process by thermal catalytic cracking of the soap phase residue of palm oil neutralization. PhD Thesis, April 2015, UFPA, http://proderna.propesp.ufpa.br/ARQUIVOS/teses/Marcelo.pdf

[56] Z.-G. Guo, S.-R. Wang, Y.-Y. Zhu, Z.-Y. Luo, K.-F. Cen. Separation of acid compounds for refining biomass pyrolysis oil. *Journal of Fuel Chemistry and Technology*. 37 (1) (2009) 49–52

[57] X. Guo, S. Wang, Z. Guo, Q. Liu, Z. Luo, K. Cen. Pyrolysis characteristics of bio-oil fractions separated by molecular distillation. *Applied Energy*. 87 (9) (2010) 2892–2898

[58] Z. Guo, S. Wang, Y. Gu, G. Xu, X. Li, Z. Luo. Separation characteristics of biomass pyrolysis oil in molecular distillation. *Separation and Purification* Technology. 76 (1) (2010) 52–57

[59] E. D. Christensen, G. M. Chupka, J. Luecke, T. Smurthwaite, T. L. Alleman, K. Iisa, J. A. Franz, D. C. Elliott, R. L. McCormick. Analysis of oxygenated compounds in hydrotreated biomass fast pyrolysis oil distillate fractions. *Energy Fuels*. 25 (11) (2011) 5462–5471

[60] S.-Q. Li, Q. Yao, Y. Chi, J.-H. Yan, K.-F. Cen. Pilot-scale pyrolysis of scrap tires in a continuous rotary kiln reactor. *Industrial and Engineering Chemistry Research*. 43 (2004) 5133–5145

[61] J.-L. Zheng, Q. Wie. Improving the quality of fast pyrolysis bio-oil by reduced pressure distillation. *Biomass and Bioenergy*. 35 (2011) 1804–1810

[62] X.-S. Zhang, G.-X. Yang, H. Jiang, W.-J. Liu, H.-S. Ding. Mass production of chemicals from biomass-derived oil by directly atmospheric distillation coupled with co-pyrolysis. *Scientific Reports*. 3 (2013) 1120

[63] A. K. Sarma, D. Konwer. Feasibility studies for conventional refinery distillation with a (1:1) w/w of a biocrude blend with petroleum crude oil. *Energy Fuels*. 19 (4) (2005) 1755–1758

[64] Y. Elkasabi, C. A. Mullen, A. A. Boateng. Distillation and isolation of commodity chemicals from bio-oil made by tail-gas reactive pyrolysis. *Sustainable Chemical Engineering*. 2 (2014) 2042–2052

[65] T. P. Vispute, G. W. Huber. Production of hydrogen, alkanes and polyols by aqueous phase processing of wood-derived pyrolysis oils. *Green Chemistry*. 11 (2009) 1433–1445

[66] Q. Song, J. Nie, M. Ren, Q. Guo. Effective phase separation of biomass pyrolysis oils by adding aqueous salt solutions. *Energy Fuels*. 23, (2009) 3307–3312

[67] ANP No. 65, 9.12.2011—DOU 12.12.2011

[68] ANP No. 37, 01.12.2009—DOU 02.12.2009

Chapter 8

Energy Evaluation of the Use of an Absorption Heat Pump in Water Distillation Process

Rosenberg J. Romero and Sotsil Silva-Sotelo

Additional information is available at the end of the chapter

http://dx.doi.org/10.5772/67094

Abstract

It is impossible to transform the whole energy into useful work. It is impossible to increase the processes of cultivation of foods without water. It is impossible to separate all ions and minerals from water. Most of the real processes are thermodynamically irreversible. Calculations may indicate the fraction of efficiency of a thermal process. All the above mentioned facts have led to some proposals of cycles that exchange energy in order to produce a useful effect for the society. The key sustainable related parts are: to obtain water free of minerals or ions by distillation process. In this chapter, thermodynamic cycles will be explained for distillation using thermodynamic cycles of thermal machines called absorption heat pumps (AHPs). Distillation process offers to the AHPs an opportunity to diminish the consumptions of fossil fuels. The AHPs are able to work with 2% of mechanical energy to carry out a sustainable distillation process. Most of the energy of an absorption heat pump operation is thermal energy. The operation of the AHPs are defined with the coefficient of performance (COP); the variations of this parameter are shown as function of the different scenarios to obtain sustainable distilled water.

Keywords: distillation, absorption heat pump, thermodynamic cycles

1. Introduction

One of the most useful processes for the sustainability would consider the efficient use of energy, water, and foods. Regrettably to include these three participants involves a process of consumption of those. It is impossible to transform the whole energy into useful work. It is impossible to increase the processes of cultivation of foods without water. It is impossible to separate all ions and minerals from water. The thermodynamics has identified the processes like reversible or irreversible process. Most of the real processes are irreversible. Calculations may indicate the fraction of efficiency of a thermal process. All the abovementioned facts have led to some proposals of cycles that exchange energy and matter with its surroundings, in order to produce a useful effect for the society. Excepting the production of foods, both key sustainable parts are related: to obtain water free of minerals or ions, it can be carried out with distillation process. In this chapter, thermodynamic cycles will be explained for distillation using thermodynamic cycles of thermal machines called absorption heat pumps (AHPs).

The main advantage of an absorption heat pump is that almost the whole energy that manipulates is thermal. Opposing to the conventional water purification systems, smaller energy is required to carry out the purification. The biggest difference with the heat exchanger is that it carries out the purification of water at different thermal levels. In the case of a heat exchanger, the energy source, where the energy comes, is at a higher thermal level (see **Figure 1**).

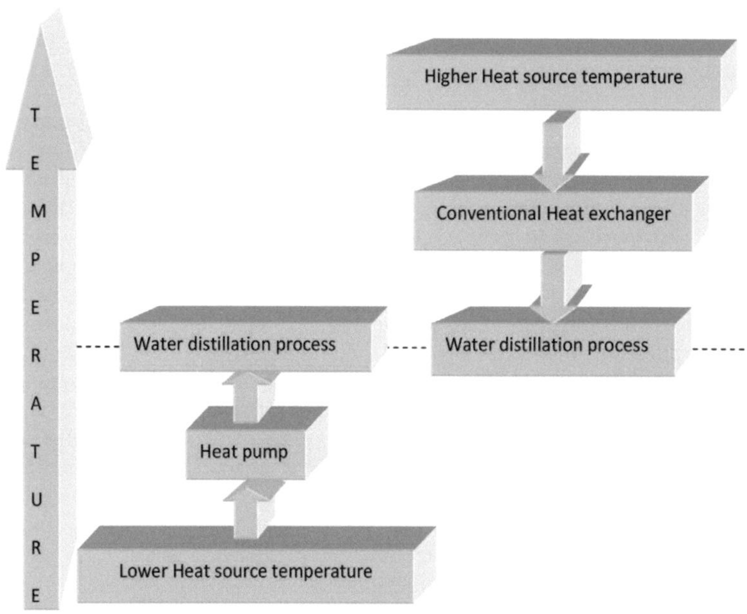

Figure 1. Thermal levels for water distillation.

2. Absorption heat pump

An absorption heat pump is a physic-chemical device made for use of sustainable thermal energy.

One of the obstacles of the technology is the ignorance that one has of them. The heat pumps are a type of thermal machines with subtle popularization. The basic concept is an analogy

http://dx.doi.org/10.5772/67094

with a mechanical pump for water. A mechanical pump for water drains water from a lower level, and by application of mechanical energy, the water goes to a higher level. In similar way, a heat pump takes energy at a lower level and sends it to a higher level, using a part of mechanical energy. This process actually happens, and it is called compression heat pumps (CHPs). If same process happens with thermal energy, then it is called absorption heat pump. The energy may go into a heat pump from several kinds of sources:

- Geothermal [1–7],
- Solar [8–11],
- Natural gas [12–14],
- Air [15–19],
- Groundwater [20–22]
- Even from mechanical energy [23, 24]
- Coal process [25–27]

The next section details the characteristics for compression and absorption heat pumps.

2.1. Compression heat pumps

A mechanical vapor compression heat pump (CHP) is a mechanical device that works with mechanical energy. The process that happens into a CHP is a thermodynamic cycle with two operations: compression and expansion (see **Figure 2**). CHP needs a fluid with the ability to change from liquid to gas. The operating conditions are constants for pressure and temperature all the time. Required energy for these changes is mechanical energy. Energy comes from lower temperature energy to the evaporator at the lowest pressure in the cycle. The delivering energy from CHP to the surroundings (i.e., for water distillation) goes at the highest pressure in the condenser. This process is a cycle, and it requires intermittent or constant energy [28].

2.2. Absorption heat pump

An absorption heat pump (AHP) is a mechanical device that does same function as CHP. It substitutes the compressor by two additional thermal components. The permanent components for a heat pump are a condenser and an evaporator. The additional thermal components which turn a CHP into an AHP are a vapor generator or called desorber and vapor absorber.

The AHP is classified in a simple way [29]: Type I if condenser temperature is higher than evaporator temperature ($T_{CO} > T_{EV}$) and Type II if condenser temperature is lower than evaporator temperature ($T_{CO} < T_{EV}$).

Type I AHP is called "conventional heat pumps," and Type II AHP is called "inverse heat pumps" or "absorption heat transformers."

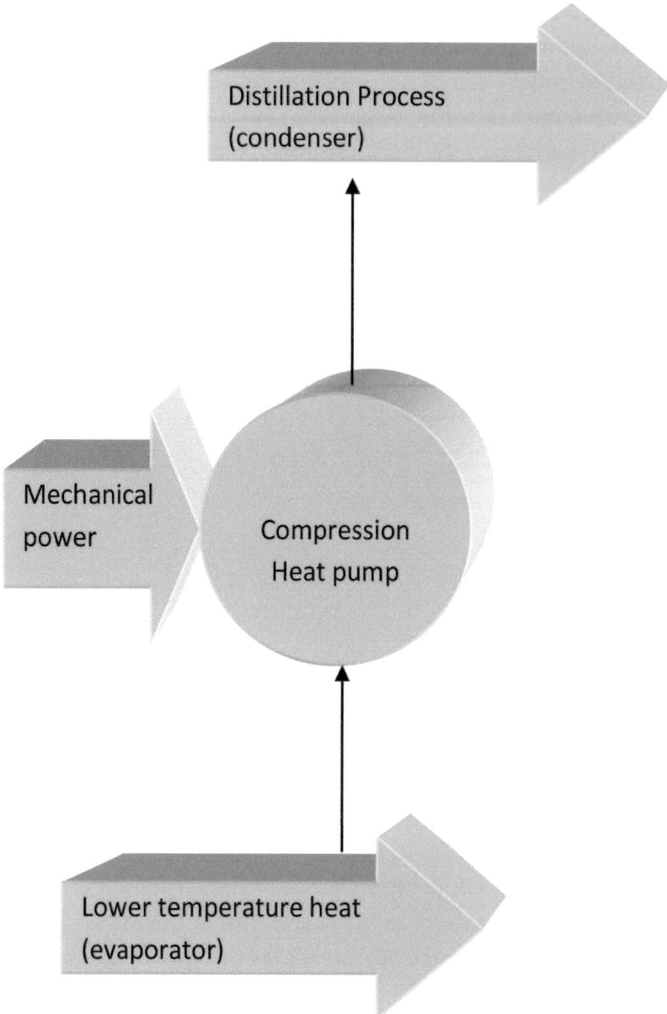

Figure 2. Schematic diagram for compression heat pump.

2.2.1. Type I heat pumps

For heat pumping, an AHP really does a useful energy transference from a lower temperature-level source of energy, free of cost (i.e., air, water, or soil) to a higher temperature level. The cycle is divided into two parts: the first part is a "working fluid" desorption from liquid absorbent at the highest temperature in the system. The second part happens at lower temperature and lower pressure for easy energy transference. When "working fluid" goes in vapor phase with liquid absorbent (separated in the first part), then a heat delivering at intermediate temperature occurs. When vapor is condensed at the highest pressure, it delivers heat at intermediate temperature also. These processes may be possible to obtain almost two thermal

http://dx.doi.org/10.5772/67094

energy units for each thermal energy unit with cost, because the energy with no cost from air, water, or soil is able permanently.

Figure 3 shows schematically these processes. There are three thermal levels in a Type I AHP. **Figure 4** shows the pressure zones and the thermal levels for the particular processes into the thermodynamic cycle.

The high pressure zone is useful for thermodynamic explanation. For a "pair" ("working fluid" and "absorbent") like water-lithium bromide, the vapor generation happens at

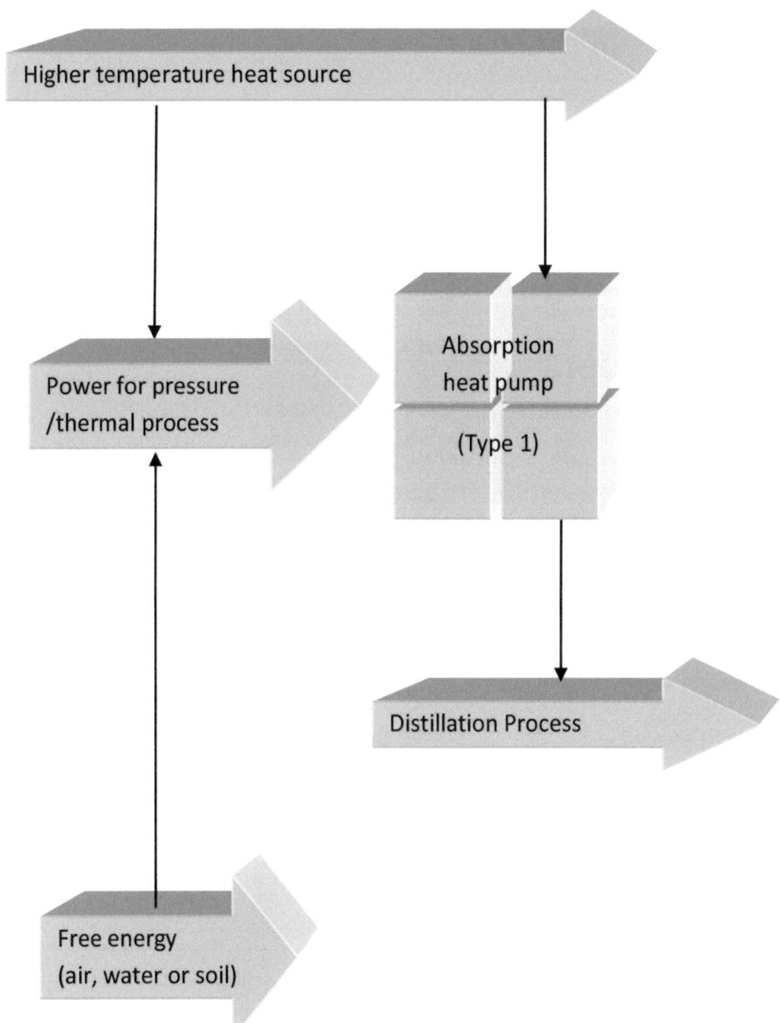

Figure 3. Absorption heat pump (Type 1) schematic concept.

temperatures lower than 100°C because there is vacuum pressure, and then the concept of relative pressure is essential.

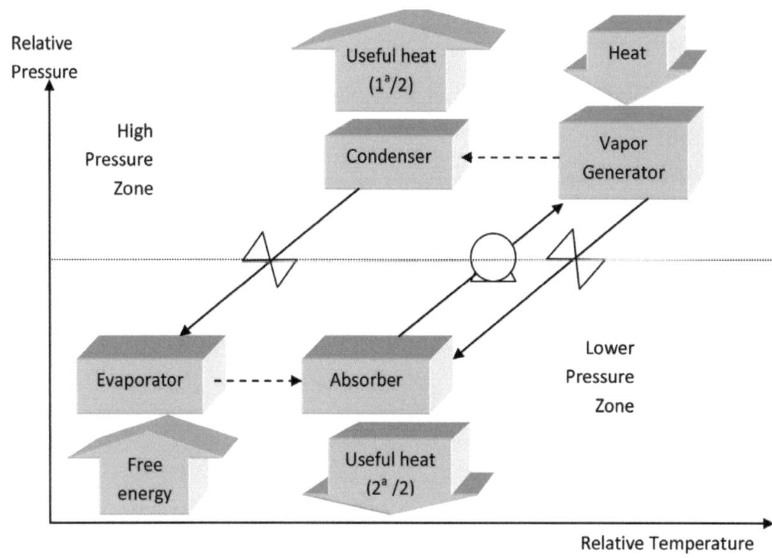

Figure 4. Absorption heat pump schematic cycle.

The thermodynamic cycle starts with energy going to "vapor generator" at higher temperature and higher relative pressure. The "working fluid" is separated from "absorbent," and it condensates in condenser unit. The condensation process releases useful energy but at lower energy level than "vapor generator." The liquid from condenser is expanded to the evaporator. In evaporator, "working fluid" is placed in contact with a free cost of energy source (air, water, or soil) at the lowest temperature level of the cycle. This temperature defines the pressure value of the entire lower pressure zone. The evaporator generates vaporized working fluid a low temperature by low-pressure condition while the absorbent with diminished working fluid mass concentration is expanded from the "generator" to the absorber. The liquid that is coming from the generator is placed with the vaporized working fluid into the absorber, and this process delivers useful energy at intermediate temperature. The result for this cycle is that there are two components delivering useful heat for just one energy entry. The solution with absorbent and working fluid is pumping to the generator to restart the cycle. The energy balance is obvious:

$$Q_{GE} + Q_{EV} = Q_{CO} + Q_{AB} \tag{1}$$

where Q_{GE} is the heat at the highest relative temperature entering to the "vapor generator" in the AHP, Q_{EV} is the heat at the lowest temperature level and coming from free cost of energy source, Q_{CO} is the useful energy at intermediate temperature caused by latent heat from vaporized "working fluid" originated in the generator, and Q_{AB} is the useful heat delivering for the junction of absorbent with working fluid at lower relative pressure.

http://dx.doi.org/10.5772/67094

The thermal efficiency for thermal machines is based on "coefficient of performance" (COP). This COP is defined for each system. For this AHP, the COP is defined as the ratio of useful energy by energy with cost. In this AHP process, the pumping work is negligible because that is around 1% of the thermal energy:

$$COP_I = \frac{Q_{CO} + Q_{AB}}{Q_{GE}} \tag{2}$$

This dimensionless value allows comparison with other technologies even with different power capacities, higher or lower. A COP value equal to 1 means all entering energy in the AHP is useful. Values higher than 1 mean that cost-free energy is added to useful energy.

2.2.2. *Type II absorption heat pumps*

The main difference between Type I and Type II absorption heat pumps (AHP) is basically the switch between the pressure zones. **Figure 5** shows the pressure and temperature levels to note that difference in the operation modes.

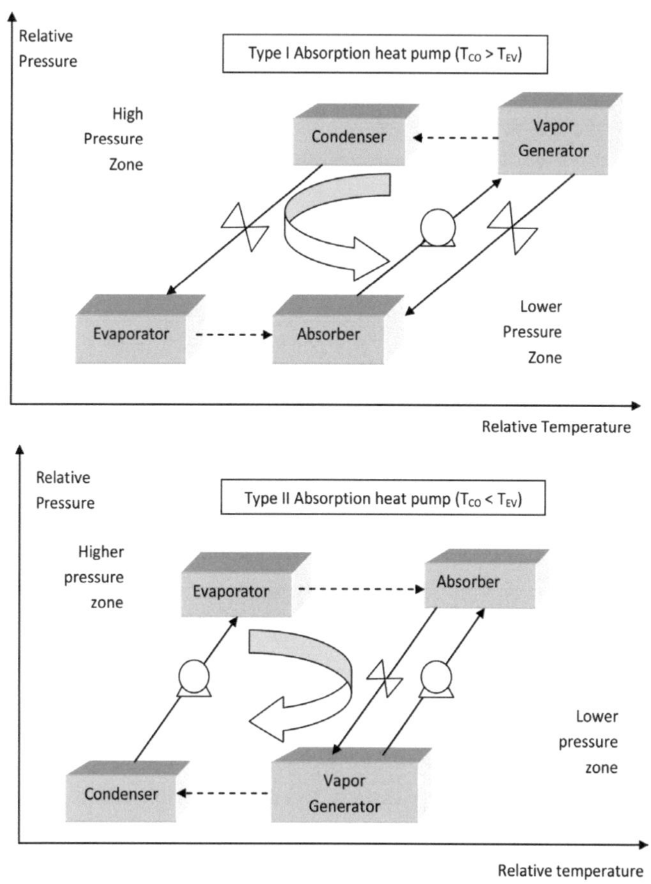

Figure 5. Type I and Type II absorption heat pumps.

Unit operations for both AHP types are the same, but they happen at different pressure and temperature values. So, the mass balance is same for Type I and Type II AHP.

The cycle occurs in similar way (see **Figure 6**). Thermal energy enters to generator and evaporator. Thermal energy is delivered in absorber and condenser. The main advantage is that energy is added at intermediate temperature level, and useful energy is delivered at higher energy level but only in absorber component.

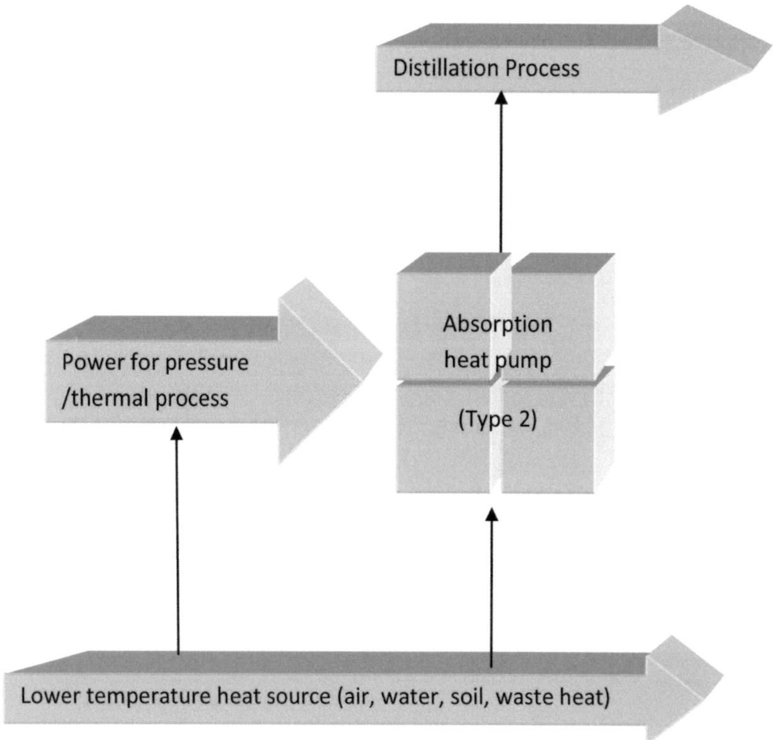

Figure 6. Absorption heat pump (Type 2) schematic concept.

This process has a particular implication: this is the unique technology that transforms thermal energy to mechanical pressure lift to deliver energy at higher temperature than the energy source.

Cycle starts when energy enters to the generator and evaporator. The generator splits working fluid in vapor phase from absorbent. The working fluid is pumping through the evaporator. Evaporator is at higher relative pressure. Evaporator changes to vapor phase from the working fluid. This vapor is contacting with liquid absorbent into the absorber. The liquid absorbent comes from the generator at intermediate temperature. The absorption process delivers energy at the highest pressure in the system. The diluted solution returns to the generator to restart the cycle.

http://dx.doi.org/10.5772/67094

COP for Type II AHP, with negligible 2 % of pumping work, is

$$COP_{II} = \frac{Q_{AB}}{Q_{GE} + Q_{EV}} \quad (3)$$

This dimensionless value is lower than the unit because the utilized energy is used as part for pressure lift and other energy parts are revalorized. It is notorious to think that if energy comes from renewable energy or waste heat [30], then the results of the CO_2 emission diminish.

3. Distillation process

Distillation process is expensive energetically. This process has two parts: the first part is to rise the liquid's temperature from an actual condition to saturation point, and the second part is to add a lot of energy to phase change process. This process has two processes, and every part has inefficiency depending on heat transfer process into vapor phase or gaseous phase.

Gases exhibit low heat transfer coefficients for the nature of molecular composition. The liquid's molecules have higher thermal conductivity than gases' molecules at same pressure and temperature conditions. Then, it is a better strategy to design energy transfer process with bigger transfer areas for liquid falling films to promote distillation process.

3.1. Water distillation

Water distillation with heat pump is no new [31–33]. However, it has lower reports [34]. The citation to papers indicates that some systems have been implemented for that purpose. The installation has two main problems: low efficiency for comparison with other technologies [35–39] and higher cost compared with conventional systems [40].

Type II AHPs are called "absorption heat transformers" also, and those have been reported for water distillation evaluation [41, 42]. Some configuration has success like "solar ponds" coupled to "absorption heat transformer" [43–45] and solar systems [46] among others renewable energies [33].

The main objective for this technology is to obtain mineral-free water with lower ambient impact [47, 48].

4. Water distillation with absorption heat pumps

Water distillation from seawater or wells may be realized with AHP with a great advantage: lower ambient emission [32, 49, 50] in order to optimize the water distillation with the minimal energy supply [51].

The modeling for this process has been previously reported with author's collaborators (i.e., [52]) that is based on mass and energy balances. It has assumptions and considerations for iterative calculation at steady-state conditions. The assumptions are close to reality to identify potential operating conditions for water distillation with Type I and Type II AHPs, with a huge amount of waste heat or renewable lower temperature heat.

4.1. Operating conditions

Water distillation requires a constant higher thermal energy. This temperature is a function of sea-level altitude at atmospheric pressure. It is not recommendable to include a vacuum process or artificial pressure increase for water distillation because those increase the energy consumption for mineral separation from distilled water [53]. For this reason, this chapter shows pairs based on water-lithium bromide for water distillation purposes for both AHP modes: Type I and Type II with:

- Lithium bromide-water
- Lithium bromide-ethylene glycol-water (called Carrol [29])

There is a possibility to use advanced AHP (double absorption or double stage [33]), but those configurations have lower COP values, and then they are not shown in the next results. Operating conditions are calculated for water distillation at a sea-level atmospheric pressure.

5. Results

The results are based on dimensionless COP that means how many energy is used for water distillation as function of each thermal energy unit added to the AHP. Of course, this dimensionless COP allows the comparison between the AHP systems.

Figure 7 shows the ratio of energy used for distillation of water with a Type I AHP using water-lithium bromide. It is clear that all energy is not useful. Some part is a waste into the cycle, while evaporator temperature is closed to ambient temperature. The best operating conditions for this distillation are for waste heat at 160°C, with a preheating treatment for water saturation at 80°C and distillation into the absorber component at 105°C.

Figure 8 shows the dimensionless COP for comparison with water-lithium bromide-ethylene glycol as additive to avoid risk in crystallization for lithium bromide. The additive has no variation in the used energy for water distillation. The dimensionless value is almost same for water distillation process at same operating conditions, but there is a variation of 10% away from risk into actual operation for leak of fracture in tubes.

Figure 9 shows the proposal of Type II AHP for water distillation using water-lithium bromide. This is a great opportunity to operate a cycle with revalorization of waste energy. The dimensionless COP value is lower than of Type I, as expected. But the temperature values are lower. Water distillation happens with only 80°C heat source, at 20°C in the ambient temperature. The thermal efficiency goes from 0.38 to 0.48 for waste heat. This is a notorious result. It is not necessary to raise energy at 100°C to water distillation. This technology allows distillation for well water with renewable energy lower than conventional machines.

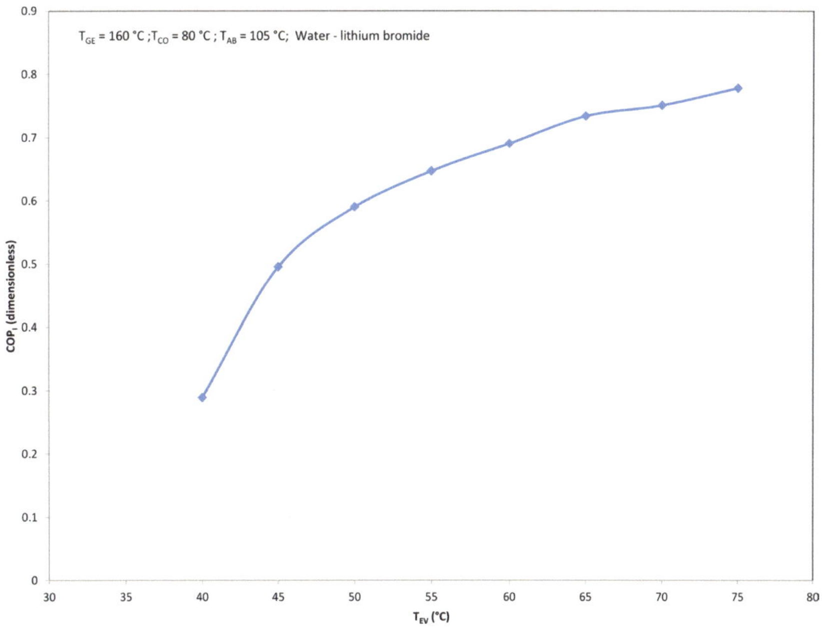

Figure 7. Dimensionless COP Type I AHP for water distillation.

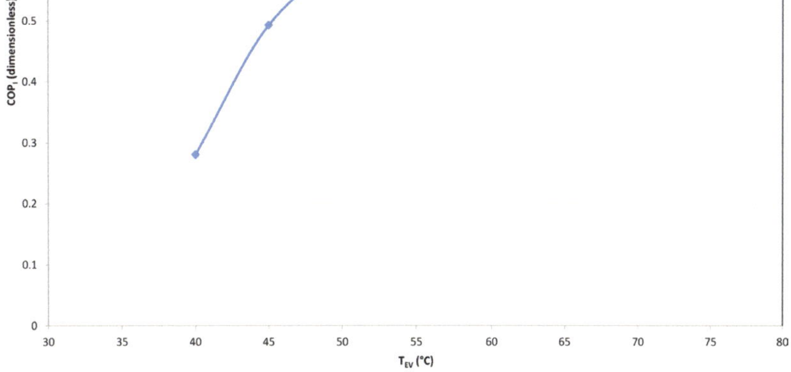

Figure 8. Dimensionless COP Type I AHP for water distillation.

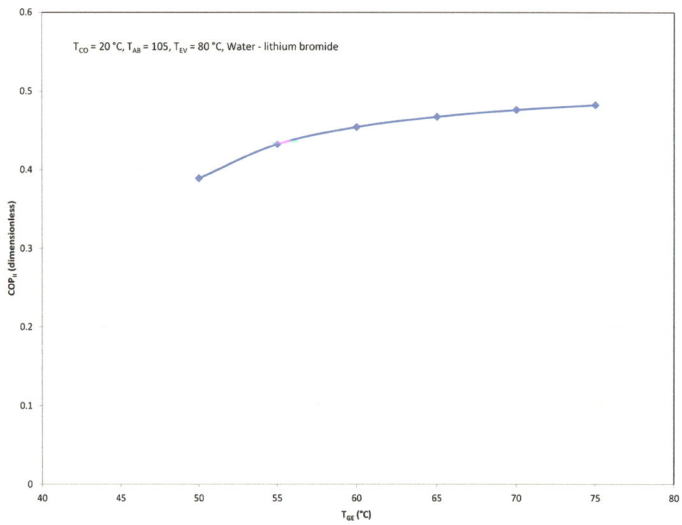

Figure 9. Dimensionless COP Type II AHP for water distillation operating with water-lithium bromide.

Finally, **Figure 10** shows a variation of the last operation conditions. The additive ethylene glycol is used to avoid crystallization. There is no variation in thermodynamic conditions for this distillation purpose, but safety for risk in crystallization is obvious. This actual operation avoids risk around 90% compared to lithium bromide.

The distillation is possible with AHP with thermal efficiencies from 0.38 to 0.78 with no fossil fuels using water-lithium bromide-ethylene glycol as pair.

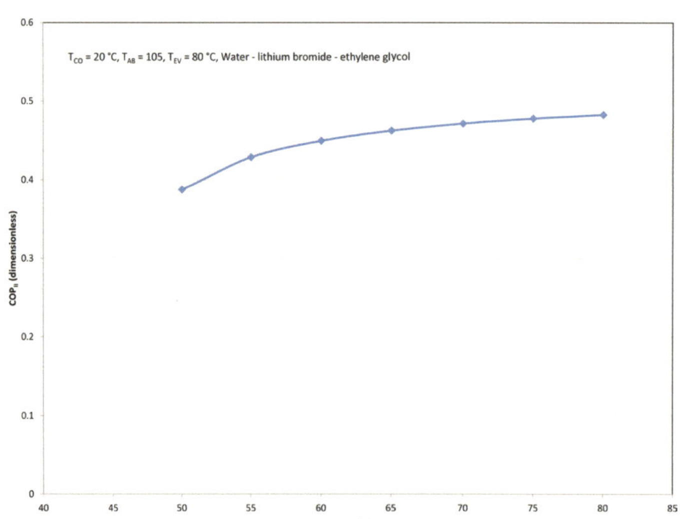

Figure 10. Dimensionless COP Type II AHP for water distillation operating with water – Carrol.

http://dx.doi.org/10.5772/67094

6. Conclusion

The energy-efficient policies with renewable energy integration are not an actual trend; this is a requirement for sustainability. Renewable energies coupled to thermodynamic cycles for water distillation have been reported previously for another authors with energy input of PV + RO around 3–20 kWh/m^3 and PV + WIND + RO around 3–16 kWh/m^3.

Conventional mechanical vapor compression plants have energy input around 11–14 kWh/m^3 for commercial purposes. Some energy for that process comes from fossil fuel with GWP gas emissions.

For absorption heat pumps, the efficient energy use is obvious: the thermal energy for these cycles (Type I and Type II) comes from waste energy or renewable energy; then there are no CO_2 emissions, while these are operated and provide distilled water.

The cycles in this chapter show water distillation at atmospheric pressure with heat exchange at 100°C with two stages: one in condenser unit to preheat the water and a second unit, the absorber, for flash process. This combination allows a distilled energy at 3.75 kWh/m^3 for Type I absorption heat pump and 8.5 kWh/m^3 for Type II absorption heat pump. These values include the actual COP with a higher distilled water production at the higher evaporator temperature in both types.

Acknowledgements

The authors thank CB-167434 Basic Science Project for the partial support.

Author details

Rosenberg J. Romero* and Sotsil Silva-Sotelo

*Address all correspondence to: rosenberg@uaem.mx

Engineering and Applied Science Research Center (CIICAp), Basic and Applied Science Institute (IICBA), Autonomous University of Morelos State (UAEM), Cuernavaca, Morelos, Mexico

References

[1] Islam, M. M., Mun, H. S., Bostami, A. R., Ahmed, S. T., Park, K. J., & Yang, C. J. (2016). Evaluation of a ground source geothermal heat pump to save energy and reduce CO_2 and noxious gas emissions in a pig house, Energy and Buildings, 111, 446-454.

[2] Verda, V., Cosentino, S., Russo, S. L., & Sciacovelli, A. (2016). Second law analysis of horizontal geothermal heat pump systems, Energy and Buildings, 124, 236-240.

[3] Staiger, R., & Tanţău, A. (2015). Energy efficiency model for small/medium geothermal heat pump systems, Management & Marketing, 10(1), 12-33.

[4] Gehlin, S. E., Spitler, J. D., & Hellström, G. (2016, January). Deep boreholes for ground source heat pump systems—Scandinavian experience and future prospects. In ASHRAE Winter Meeting, Orlando, Florida (pp. 23-27).

[5] Go, G. H., Lee, S. R., Yoon, S., & Kim, M. J. (2016). Optimum design of horizontal ground-coupled heat pump systems using spiral-coil-loop heat exchangers, Applied Energy, 162, 330-345.

[6] Liu, Z., Xu, W., Qian, C., Chen, X., & Jin, G. (2015). Investigation on the feasibility and performance of ground source heat pump (GSHP) in three cities in cold climate zone, China, Renewable Energy, 84, 89-96.

[7] Bleicher, A., & Gross, M. (2016). Geothermal heat pumps and the vagaries of subterranean geology: Energy independence at a household level as a real world experiment, Renewable and Sustainable Energy Reviews, 64, 279-288.

[8] Gurin, M. H. (2016). U.S. Patent No. 9,316,404. Washington, DC: U.S. Patent and Trademark Office.

[9] Zhao, M., Gu, Z. L., Kang, W. B., Liu, X., Zhang, L. Y., Jin, L. W., & Zhang, Q. L. (2017). Experimental investigation and feasibility analysis on a capillary radiant heating system based on solar and air source heat pump dual heat source, Applied Energy, 185, 2094-2105.

[10] Yan, G., Bai, T., & Yu, J. (2016). Energy and exergy efficiency analysis of solar driven ejector-compressor heat pump cycle, Solar Energy, 125, 243-255.

[11] Yang, W., Sun, L., & Chen, Y. (2015). Experimental investigations of the performance of a solar-ground source heat pump system operated in heating modes, Energy and Buildings, 89, 97-111.

[12] Hong, B., & Howarth, R. W. (2016). Greenhouse gas emissions from domestic hot water: heat pumps compared to most commonly used systems, Energy Science & Engineering, 4, 123-133.

[13] Ghezelbash, R., Farzaneh-Gord, M., Behi, H., Sadi, M., & Khorramabady, H. S. (2015). Performance assessment of a natural gas expansion plant integrated with a vertical ground-coupled heat pump, Energy, 93, 2503-2517.

[14] Wei, M., Yuan, W., Song, Z., Fu, L., & Zhang, S. (2015). Simulation of a heat pump system for total heat recovery from flue gas, Applied Thermal Engineering, 86, 326-332.

[15] Ninikas, K., Hytiris, N., Emmanuel, R., Aaen, B., & Younger, P. L. (2016). Heat recovery from air in underground transport tunnels, Renewable Energy, 96, 843-849.

[16] Yıldız, İ., Yue, J., & Yıldız, A. C. (2015). Air source heat pump performance in open, semi-closed, and closed greenhouse systems in the Canadian prairies, In Progress in Clean Energy, Volume 1 (pp. 203-212). Springer International Publishing, Switzerland.

[17] Yıldız, İ., Yue, J., & Yıldız, A. C. (2015). Air source heat pump performance in open, semi-closed, and closed greenhouse systems in Ontario, In Progress in Clean Energy, Volume 1 (pp. 193-202). Springer International Publishing, Switzerland.

[18] Yıldız, İ., Yue, J., & Yıldız, A. C. (2015). Air source heat pump performance in open, semi-closed, and closed greenhouse systems in Quebec and Labrador, In Progress in Clean Energy, Volume 1 (pp. 213-223). Springer International Publishing, Switzerland.

[19] Li, Y., & Yu, J. (2016). Theoretical analysis on optimal configurations of heat exchanger and compressor in a two-stage compression air source heat pump system, Applied Thermal Engineering, 96, 682-689.

[20] Wu, Q., Xu, S., Zhou, W., & LaMoreaux, J. (2015). Hydrogeology and design of groundwater heat pump systems, Environmental Earth Sciences, 73(7), 3683-3695.

[21] Park, Y., Kim, N., & Lee, J. Y. (2015). Geochemical properties of groundwater affected by open loop geothermal heat pump systems in Korea, Geosciences Journal, 19(3), 515-526.

[22] Zheng, T., Shao, H., Schelenz, S., Hein, P., Vienken, T., Pang, Z, & Nagel, T. (2016). Efficiency and economic analysis of utilizing latent heat from groundwater freezing in the context of borehole heat exchanger coupled ground source heat pump systems, Applied Thermal Engineering, 105, 314-326.

[23] Shah, N. N., Huang, M. J., & Hewitt, N. J. (2016). Experimental study of a diesel engine heat pump in heating mode for domestic retrofit application, Applied Thermal Engineering, 98, 522-531.

[24] Li, Y., Yan, M., Zhang, L., Chen, G., Cui, L., Song, Z., & Ma, C. (2016). Method of flash evaporation and condensation-heat pump for deep cooling of coal-fired power plant flue gas: Latent heat and water recovery, Applied Energy, 172, 107-117.

[25] Athresh, A. P., Al-Habaibeh, A., & Parker, K. (2016). The design and evaluation of an open loop ground source heat pump operating in an ochre-rich coal mine water environment, International Journal of Coal Geology, 164, 69-76.

[26] Zhang, H., Li, Z., & Zhao, H. (2015). Thermodynamic performance analysis of a novel electricity-heating cogeneration system (EHCS) based on absorption heat pump applied in the coal-fired power plant, Energy Conversion and Management, 105, 1125-1137.

[27] Xi, C., Hongxing, Y., Lin, L., Jinggang, W., & Wei, L. (2011). Experimental studies on a ground coupled heat pump with solar thermal collectors for space heating, Energy, 36(8), 5292-5300.

[28] Herold, K. E., Radermacher, R., & Klein, S. A. (2016). Absorption Chillers and Heat Pumps. CRC Press, N.Y., U.S.A.

[29] Romero, R. J., et al. (2011). Comparison of double stage heat transformer with double absorption heat transformer operating with Carrol-Water for industrial waste heat recovery, Chemical Engineering Transactions, 25, 129-134.

[30] Jana, A. K. (2014). Advances in heat pump assisted distillation column: a review, Energy Conversion and Management, 77, 287-297.

[31] Donnellan, P., Cronin, K., & Byrne, E. (2015). Recycling waste heat energy using vapour absorption heat transformers: a review, Renewable and Sustainable Energy Reviews, 42, 1290-1304.

[32] Rivera, W., Best, R., Cardoso, M. J., & Romero, R. J. (2015). A review of absorption heat transformers, Applied Thermal Engineering, 91, 654-670.

[33] Attia, A. A. (2016). Heat pump seawater distillation system using passive vacuum generation system, Desalination, 397, 151-156.

[34] Alarcón-Padilla, D. C., García-Rodríguez, L., & Blanco-Gálvez, J. (2007). Assessment of an absorption heat pump coupled to a multi-effect distillation unit within AQUASOL project, Desalination, 212(1), 303-310.

[35] Siqueiros, J., & Romero, R. J. (2007). Increase of COP for heat transformer in water purification systems. Part I—Increasing heat source temperature, Applied Thermal Engineering, 27(5), 1043-1053.

[36] Sekar, S., & Saravanan, R. (2011). Experimental studies on absorption heat transformer coupled distillation system, Desalination, 274(1), 292-301.

[37] Khamooshi, M., Parham, K., Egelioglu, F., Yari, M., & Salati, H. (2014). Simulation and optimization of novel configurations of triple absorption heat transformers integrated to a water desalination system, Desalination, 348, 39-48.

[38] Amin, Z. M., & Hawlader, M. N. A. (2015). Analysis of solar desalination system using heat pump, Renewable Energy, 74, 116-123.

[39] Díez, E., Langston, P., Ovejero, G., & Romero, M. D. (2009). Economic feasibility of heat pumps in distillation to reduce energy use, Applied Thermal Engineering, 29(5), 1216-1223.

[40] Ibarra-Bahena, J., Romero, R. J., Velazquez-Avelar, L., Valdez-Morales, C. V., & Galindo-Luna, Y. R. (2015). Experimental thermodynamic evaluation for a single stage heat transformer prototype build with commercial PHEs, Applied Thermal Engineering, 75, 1262-1270.

[41] Parham, K., Khamooshi, M., Tematio, D. B. K., Yari, M., & Atikol, U. (2014). Absorption heat transformers—a comprehensive review, Renewable and Sustainable Energy Reviews, 34, 430-452.

[42] Rivera, W., & Romero, R. J. (2000). Evaluation of a heat transformer powered by a solar pond, Solar Energy Materials and Solar Cells, 63(4), 413-422.

[43] Rivera, W., Cardoso, M. J., & Romero, R. J. (2001). Single-stage and advanced absorption heat transformers operating with lithium bromide mixtures used to increase solar pond's temperature, Solar Energy Materials and Solar Cells, 70(3), 321-333.

[44] Salata, F., & Coppi, M. (2014). A first approach study on the desalination of sea water using heat transformers powered by solar ponds, Applied Energy, 136, 611-618.

[45] Ibarra-Bahena, J., Dehesa-Carrasco, U., Montiel-González, M., Romero, R. J., & VenegasReyes, E. (2017). Feasibility analysis of a hot water solar system coupled to an absorption heat transformer, Applied Thermal Engineering (In press, Corrected Proof).

[46] Romero, R. J., Silva-Sotelo, S., Martínez, R., & Román, J. C. (2013). Energy saving in advanced absorption heat pump with object oriented programming, In Emerging Trends in Computing, Informatics, Systems Sciences, and Engineering (pp. 1101-1111). New York: Springer.

[47] Patiño, J. A. D., Martínez, A. R., Romero, R. J., Ibarra-Bahena, J., & Patiño, M. L. D. (2016). Environmental impact assessment for an absorption heat transformer, Open Journal of Applied Sciences, 6(07), 409.

[48] Cho, H., Smith, A. D., & Mago, P. (2014). Combined cooling, heating and power: a review of performance improvement and optimization, Applied Energy, 136, 168-185.

[49] Kannan, N., & Vakeesan, D. (2016). Solar energy for future world: a review, Renewable and Sustainable Energy Reviews, 62, 1092-1105.

[50] Romero, R. J., & Rodríguez-Martínez, A. (2008). Optimal water purification using low grade waste heat in an absorption heat transformer, Desalination, 220(1), 506-513.

[51] Best, R., Rivera, W., Cardoso, M. J., Romero, R. J., & Holland, F. A. (1997). Modelling of single-stage and advanced absorption heat transformers operating with the water/carrol mixture, Applied Thermal Engineering, 17(11), 1111-1122.

[52] Kiss, A. A., Landaeta, S. J. F., & Ferreira, C. A. I. (2012). Towards energy efficient distillation technologies—making the right choice, Energy, 47(1), 531-542.

[53] Biermann, W. J., & Relnann, R. C. (1979). Properties of the Carrol System and a Machine Design for Solar-Powered, Air-Cooled, Absorption Space Cooling, Phase I and Phase II Final Report. New York: Carrier Corporation.

Chapter 9

Solar Membrane Distillation: Use of a Helically Coiled Fiber

Adel Zrelli

Additional information is available at the end of the chapter

http://dx.doi.org/10.5772/67653

Abstract

Membrane distillation (MD) is a novel process introduced to overcome the disadvantages of the conventional distillation process. MD has gained much interest, principally, for its lower energy demand and higher rejection factors. Many configurations of the membrane module have been tested to improve the MD process. In this case, and in order to see the possibility of using the helically coiled fiber for MD coupled with solar energy, a mathematical model has been developed. This model describes the evolutions of permeate flow rate with the variation of inlet feed characteristics, solar flux, and so on. A comparison between the use of linear and helically-coiled fiber shows an enhancement of the temperature polarization coefficient about 6% for the case of the helical fiber and to have an improvement factor by 28%. For the case of the effect of solar energy on the permeate flow rate, an increase of 264% is remarked for the variation of the direct solar flux from 200 to 800 W/m^2. However, a reduction of 12% for the permeate flux is obtained, when the inlet feed concentration grows from 10 to 300 g/l.

Keywords: Vacuum membrane distillation, Solar energy, Helical fiber, Simulation

1. Introduction

Water is abundant on Earth. The majority (97.2%) consists of seawater [1]. The rest is composed of 2.2% ice caps and glaciers (unusable directly), while fresh water, which is found in lakes, rivers, and groundwater, is only 0.6% [2]. In addition, the distribution of this water is very uneven. In fact, only 10 countries share 60% of freshwater reserves while 29 countries are facing a shortage of fresh water. These countries are located in Africa and the Middle East [3]. However, these freshwater may not be potable. Allowing this globally, many countries are below the threshold shortage of drinking water. The drinking water is a major issue for development and survival for humans [4]. This shortage is a result of poor management of reserves, water pollution, and population growth. Also, this water shortage affects nearly 80% of world's population [5]. To overcome this shortage of drinking water, several solutions were presented among which are the desalination of brackish water or seawater. Mainly, the desalination technologies, used around the world, are the thermal with 33% of total capacity and membrane processes with 56% of total capacity [6]. The first type, thermal processes, is based on physical change in the water state. For this type, the two techniques, which gained great global interest to offer to yet effective solutions to desalinate water, are multistage flash distillation (MSF) and multi-effect distillation (MED) [7]. For the second type, which is the membrane processes, a membrane is used to separate the water from the saline water (brackish or seawater). In this case, the characteristics of the membrane (such as porosity, wettability, selectivity, electric charge, etc.) are so important in order to obtain a high efficiency for the membrane processes. The two main techniques, used in this type, are the reverse osmosis (RO) and the electrodialysis (ED) [6]. Each technique uses the ability of membranes to separate selectively salts and water from the saline water. These two types of desalination technologies (thermal and membrane processes) have a high-energy demand, and the process performance is limited by the osmotic pressure or concentration polarization. In order to overcome these drawbacks, a promising alternative technology for desalination is introduced, which is the membrane distillation (MD) due principally to lower energy cost and membrane fouling [8]. MD is composed of four configurations. These configurations are the direct contact membrane distillation (DCMD), the air gap membrane distillation (AGMD), the sweeping gas membrane distillation (SGMD), and the vacuum membrane distillation (VMD) [9]. VMD has attracted increasing interest for various applications. From energy consumption point of view, it could clearly compete with reverse osmosis, when coupled with alternate source of energy such as solar energy [10]. In order to couple VMD to solar energy, different configurations can be used. In this case, membrane can be placed in or out of the absorber of the solar collector. Furthermore, many configurations of the hollow fiber membrane can be used. Among these configurations, we found the linear and the helical fibers.

In this chapter, a brief outline of the main thermal and membrane desalination processes will be presented. This will be followed by the exposure of the membrane distillation and its different configurations and the used hollow fiber geometries in order to increase the process performance. Therefore, this chapter attempts to develop a novel hollow fiber module design, which is the helically coiled fiber. A comparison of the performance of the helically coiled hollow fiber to those of linear fiber will be done. The effects of this novel configuration on the permeate flow rate and the temperature polarization coefficient will be highlighted. In addition, the effects of the process conditions, such as the initial feed concentration and the solar radiation, on the permeate flow rate will be evaluated.

2. Desalination processes

The most important desalination processes are the thermal processes and membrane processes.

2.1. Thermal processes

2.1.1. Multi-effect distillation (MED)

These processes are composed principally of multistage flash (MSF) and multi-effect distillation (MED). MED is the oldest technique applied for seawater desalination [11]. The MED principle is based on heat transfer from the condensing steam to feed seawater or brine in a series of effects (**Figure 1**). In this case, the steam produced in the first effect was condensed to produce fresh water in the second effect. This latter effect was operated at slightly lower pressure and temperature than in the previous effect. The heat of steam condensation allows to evaporate a portion of the seawater contained in the second effect and so on. Thus, only the energy required for the evaporation in the first effect is of external origin. MED can work with a low steam temperature as heat source (70–80°C) and the multiplication of effects allows to reduce the specific consumption. Therefore, the number of effects is usually between 8 and 16 [12, 13].

2.1.2. Multistage flash (MSF)

The MSF process comes in practice since the late 1950s [15]. The distillation was done, in this case, in a series of flash chambers. The generation of steam, from brackish water or seawater, was the result of a pressure drop and not of heat exchange with condensing steam, which is the case of MED. The principle is so simple, the brackish water or seawater was pressurized and heated to the plant's maximum allowable temperature (limited to about 110°C). Afterwards, this heated liquid was introduced into a chamber maintained at slightly below the saturation vapor pressure of the water (**Figure 2**). This chamber pressure level induces the flash of a fraction of its water content into steam. This latter is converted into fresh water by condensation on heat exchanger with a series of closed pipes, while the rest of the heated liquid continues flowing through a series of chambers. The number of chambers, for this technology, is between 20 and 25 [11, 14, 16].

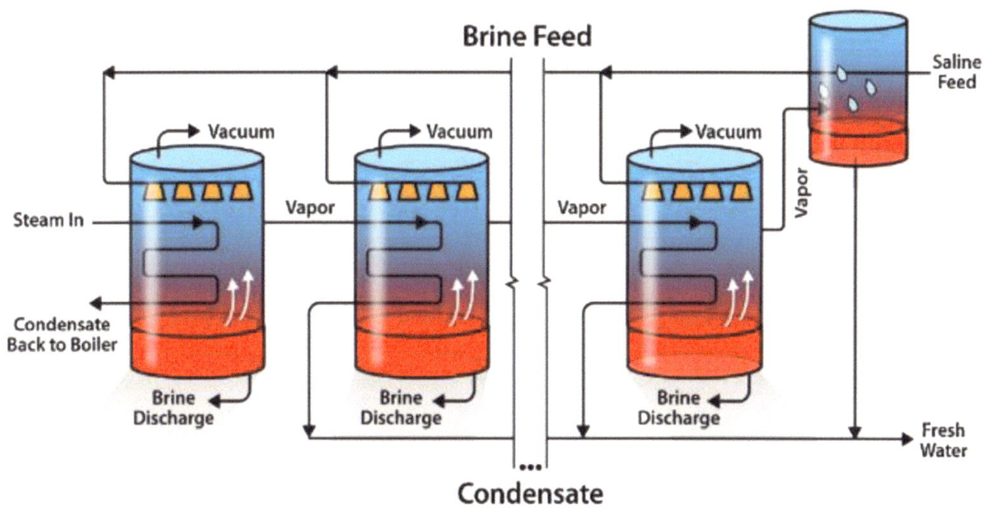

Figure 1. Diagram of a multi-effect desalination (MED) unit [14].

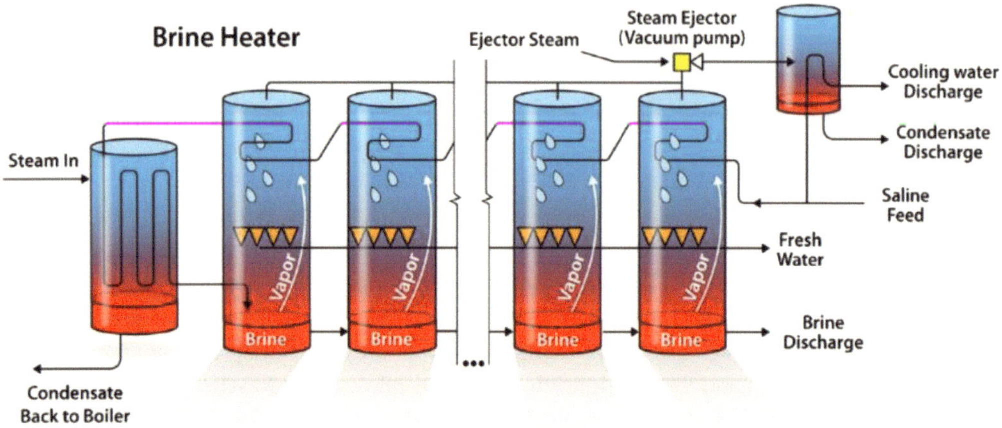

Figure 2. Diagram of a multistage flash distillation (MSF) unit [14].

2.2. Membrane processes

Membrane processes include electrodialysis (ED) and reverse osmosis (RO) [17].

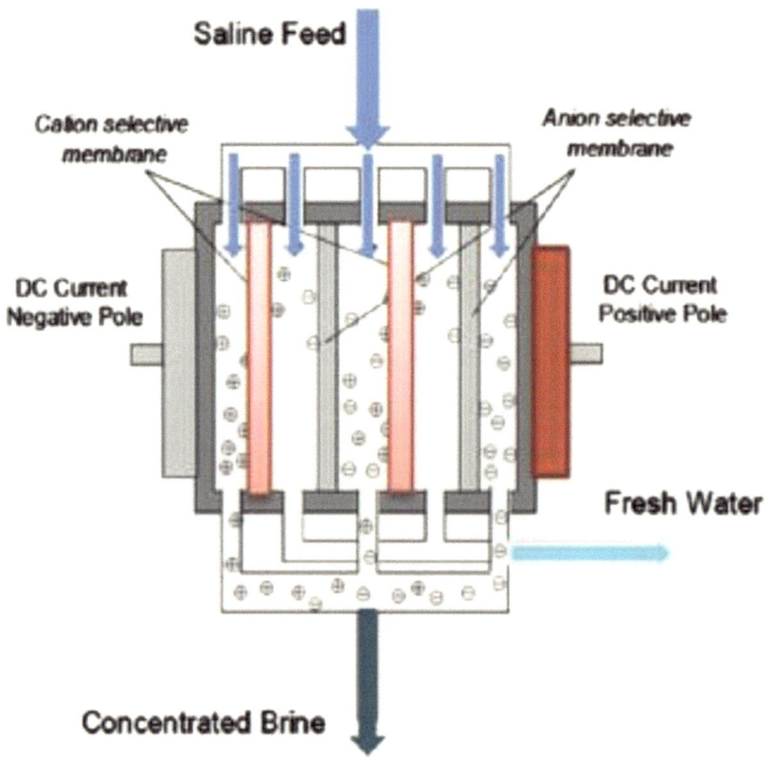

Figure 3. Schematic diagram of electrodialysis desalination process [14].

http://dx.doi.org/10.5772/67653

2.2.1. *Electrodialysis*

Electrodialysis process (ED) is an electrochemical separation process, which has been in commercial use for desalination of brackish water since 1970s, particularly for small- and medium-scale processes. This process uses the electric direct current to remove the salt ions in the brackish water. The latter passes between pairs of anion-exchange and cation-exchange membranes. The cations (positive ions) migrate from the brackish water toward the negative electrode through the cation-exchange membranes, which allow only cations to pass (**Figure 3**). On the other hand, the anions (negative ions) migrate toward the anode through the anion-exchange membranes. In a conventional process, a large number of alternating cation-exchange and anion-exchange membranes are stacked together, separated by flow spacers, which are plastic sheets that allow the passage of water. The total power consumption of ED units ranges from 0.7 to 2.5 kWh/m^3 of desalinated water for feed water salinity of 2500 ppm and from 2.64 to 5.5 kWh/m^3 of desalinated water for feed water salinity of 5000 ppm [18–20].

2.2.2. *Reverse osmosis*

Reverse osmosis (RO) is a pressure-driven process that separates two solutions with different concentrations across a semi-permeable membrane (**Figure 4**). The rate of fresh water that penetrates the membrane depends on the difference between the applied pressure and the osmotic pressure of the feed salt water. The osmotic pressure is directly related to the salt concentration in the saline water. For brackish water desalination, the operating pressures range from 18 to 28 bar, and for seawater desalination from 55 to 69 bar [21]. The discharge

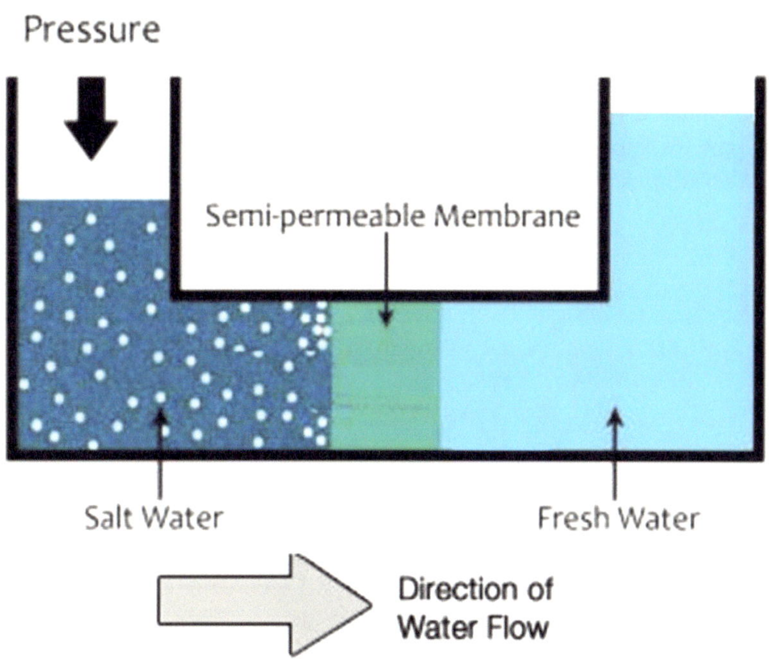

Figure 4. Schematic diagram of RO desalination process.

brine from an RO unit ranges from 20 to 70% of the flow feed water, depending on salinity of the feed water, applied pressure, and type of membrane.

Until today, the predominant desalination processes in use are RO and MSF, which constitute 53 and 25% of worldwide capacity, respectively (**Figure 5**). However, these processes have some disadvantages, which are described in the subsequent text.

They are considered energy intensive either by the heat demand for the MSF process (generally for the thermal processes) or by the high-pressure demand as in reverse osmosis. The used pressure for the RO process is about 10–15 bar for brackish water and 50–80 bar for seawater, and the consumed electrical energy, to produce 1 m^3 of desalinated water, is in the range of 3–4 kWh [23]. This high-energy consumption contributes to further environmental problems such as more pollutants and undesired emissions. Other disadvantages of these processes are the high maintenance cost of the mechanical equipment and the limited membrane lifespan [24]. Also, reverse osmosis efficiency is strongly affected by the osmotic pressure of the highly concentrate feed solutions, which imply the use of high pressure and the reduction of the salt rejection with the decay by 50% of permeate flux [25]. These drawbacks push down the efficiency of those processes, which require the search for alternative, environment friendly, and sustainable desalination.

2.3. Membrane distillation

2.3.1. Membrane distillation principle

Membrane distillation (MD) is an alternative to the traditional evaporative distillation systems used for desalination or water purification processes. MD is a thermally driven membrane process developed over the last 60 years [26]. In this process, a hydrophobic microporous membrane separates a hot and cold stream (**Figure 6**). This membrane serves, also, as a physical support for vapor transport but a barrier to liquid penetration, thus allowing the separation of volatile and non-volatile species. In the case of salty water, the water (volatile

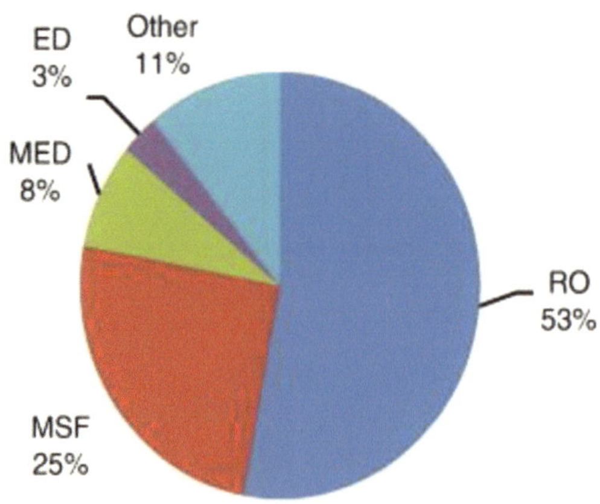

Figure 5. Distribution of plant capacity according to desalination process [6, 22].

http://dx.doi.org/10.5772/67653

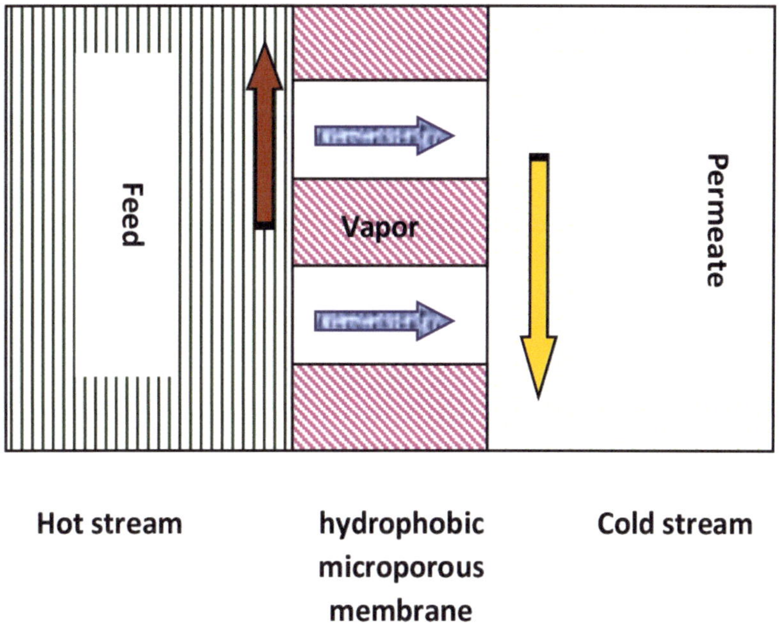

Figure 6. Principle of membrane distillation.

specie) passes through the membrane as vapor without salt and condenses on the low-temperature side and distillate is formed [27, 28].

Some of advantages of MD processes over conventional desalination processes such as MSF and RO are as follows:

- Lower working temperature, which leads to coupled MD to low-grade and renewable energy source such as solar energy.
- Lower operating pressures, which induce a less demanding membrane mechanical properties.
- Outstanding rejection performance, of non-volatile solute, which can reach as high as 100%.
- Performance is not significantly affected by high osmotic pressure or concentration polarization because the solution vapor pressure changes only marginally with salt concentration.
- Reduced chemical interaction between membrane and process solution [8, 25, 27, 29, 30].

2.3.2. MD configurations

MD can be classified into four different configurations according to the nature of the cold side of the membrane [31]: the first configuration is the direct contact membrane distillation (DCMD) in which the membrane is in direct contact with liquid phases on both sides. The volatile components of the feed evaporate at the interface feed/membrane diffuse through the membrane pores and condensate at the cold side in the distillate stream [26–29]. The second configuration is air gap membrane distillation (AGMD) in which an air gap is interposed between the membrane and the condensate surface. This stagnant air gap reduces heat losses due to conduction, thus increasing the thermal efficiency of this configuration [32]. The sweeping gas membrane distillation (SGMD) represents the third configuration. In this configuration, a cold inert gas is used in permeate side as carrier for the produced vapor. The condensation of this vapor takes place outside the membrane module [8]. The fourth and the last configuration is the vacuum membrane distillation (VMD), in which the vapor phase is vacuumed from the liquid through the membrane and condensed outside of the membrane module [30].

VMD presents many advantages when compared to the other MD configurations by presenting the highest flux and desalination rate, and the lowest fresh water conductivity (**Figures 7** and **8**). For VMD, the two main advantages are a very low conductive heat loss and a reduced mass transfer resistance [30, 33–35].

2.3.3. *Coupling MD with solar energy*

Generally, desalination consumes much energy. In order to minimize the energy usage and consequently the cost, the use of renewable energy (wind, geothermal, solar, etc.) is a feasible and potential solution. In addition, the lack of drinkable water often accompanies the abundance of solar radiation, which makes favorable the coupling of the desalination processes with the solar energy [36]. In the case of VMD, the permeate flux could be enhanced, when it is coupled with solar energy, and its value can be increased from the range of (5–15 l/h m^2) to that (40–85 l/h m^2). Following this coupling of VMD with solar energy, this MD process could clearly compete with RO [37].

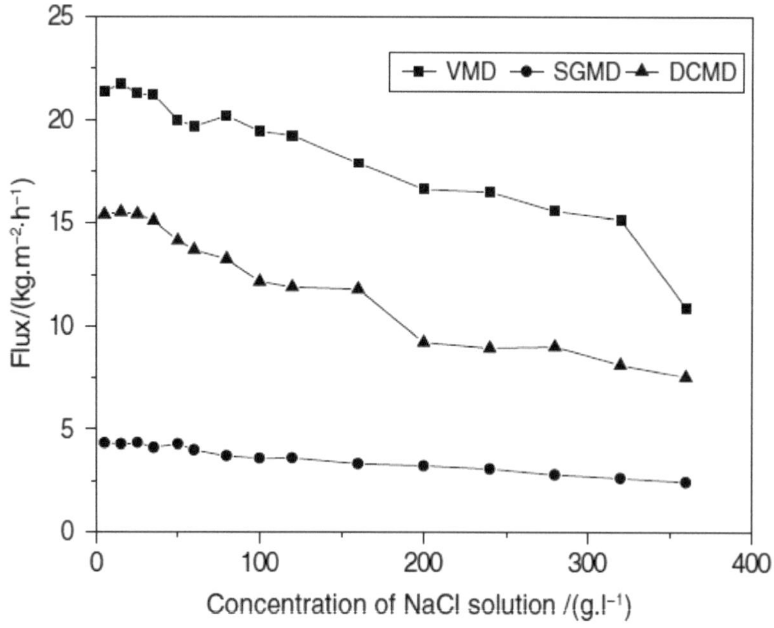

Figure 7. Comparison of flux for three configurations at different feed concentration [33].

http://dx.doi.org/10.5772/67653

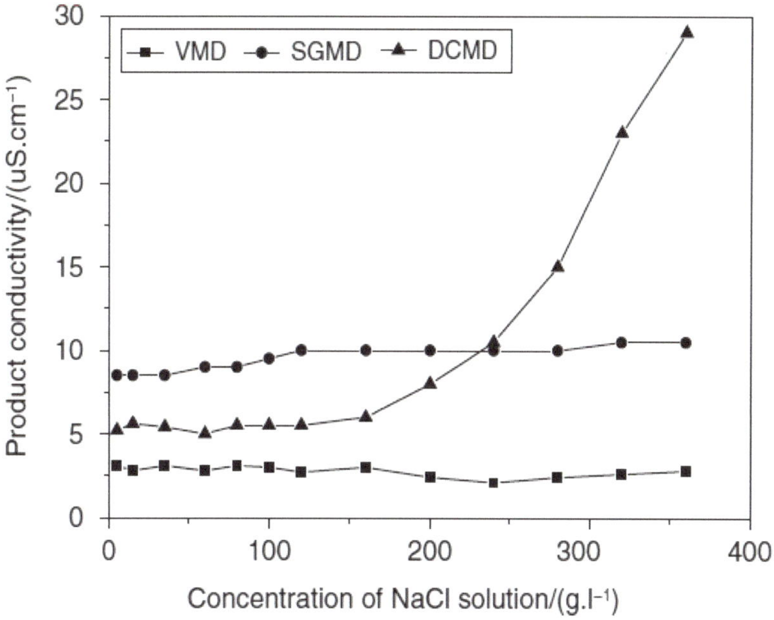

Figure 8. Comparison of conductivity for three configurations at different feed concentration [33].

2.3.4. Module geometry configurations

In MD process, the most studied geometry configurations are flat sheet and hollow fiber membrane modules. The hollow fiber membrane modules are preferable due to their larger membrane area per unit volume and highest packing density of all module types and modular versatility. The hollow fiber packing density is about 3000 m^2/m^3 [8]. However, a poor configuration of hollow fiber modules will result in the reduction of the permeate production and the efficiency of the MD process. In order to overcome this problem, many studies have focused on strategies to improve the MD performance through designing novel membrane modules [38–41] and enhancing the permeate flux. In this case, the introduction of baffles (**Figure 9**), in

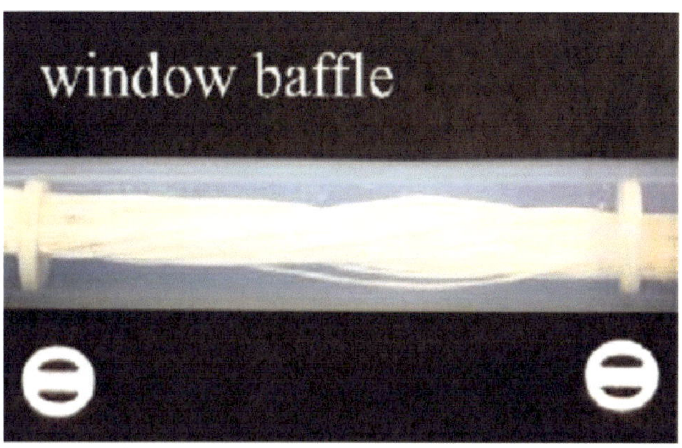

Figure 9. Hollow fiber module with window baffles [38].

membrane module, provides a better flow distribution and could increase the shell-side heat-transfer coefficients leading to 20–28% flux enhancement.

Furthermore, the use of hollow fiber configurations with wavy geometries (twisted and braided (**Figure 10**)) led to flux enhancements as high as 36%.

Yang et al. [39] reported that the greatest enhancement is achieved by the modules with spacer-knitted (**Figure 11**) for which the flux is increased more than 90% when compared with the randomly packed module. For the same configuration (spacer-knitted) and when the operating conditions were modified, the flux enhancement achieved was only 51.8% [41].

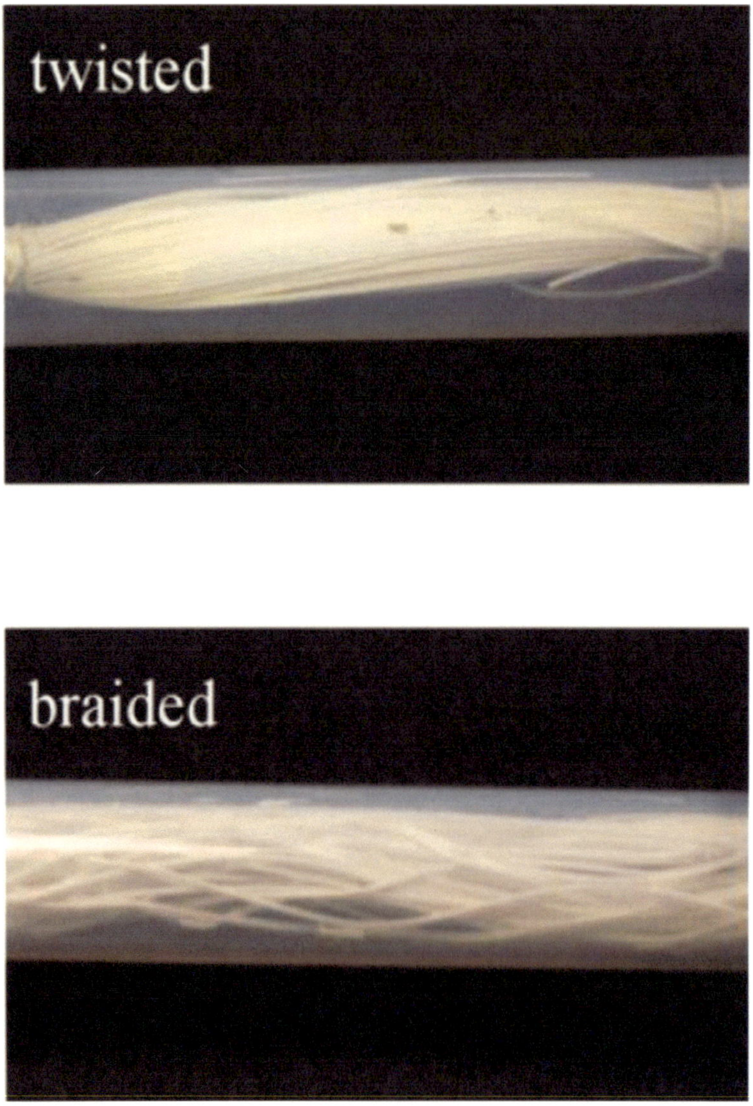

Figure 10. Wavy geometries (twisted and braided) hollow fiber module [38].

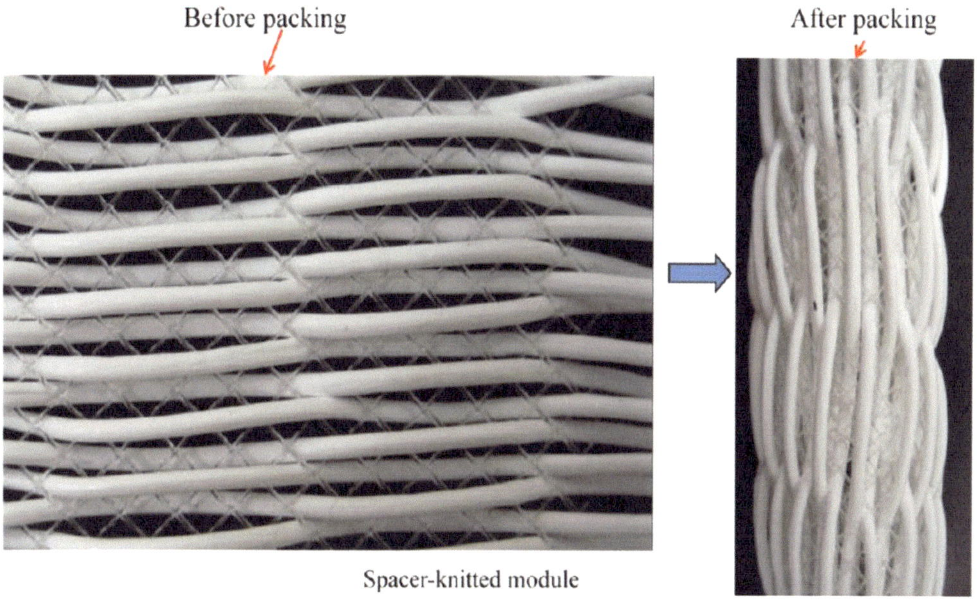

Figure 11. Spacer-knitted design of hollow fiber module [39].

On the other hand, the other fiber configuration is the helically coiled fiber. Few studies have been undertaken on the use of this configuration. Among these studies, we present the Mallubhota et al. study [42]. In this work, a comparison, for nanofiltration, between linear and helical modules was done. The results show an enhancement of the permeation rate during their use of the helical module. This enhancement is due to the reduction membrane fouling and polarization concentration. In addition, when the helical-coiled fiber was used by Nagase et al. [43], an enhancement in the mass transfer coefficient and consequently in the permeate rate has been observed. The same finding has been proved by Liu et al. when they used this helical configuration in membrane extraction.

3. Use of helically coiled fiber

3.1. Design description

In our case, the solar desalination installation (shown in **Figure 12**) is composed of parabolic trough concentrator. At the focal axis, the absorber is mounted, which is in the shape of a cylindrical tube. This absorber contains the hollow fiber membrane. This membrane has the shape of a coil and the configuration of an absorber, and the membrane is similar to a helically coiled heat exchanger. In addition to that, solar rays are focused into the absorber, and an increase of the feed temperature is consequently reported.

In order to have a symmetrical conception, we study a system that contains two helically coiled fibers (**Figure 13**). Since the flow is symmetric about a vertical plane passing through the axis of the cylinder, only the half-plane needs to be considered (**Figure 14**).

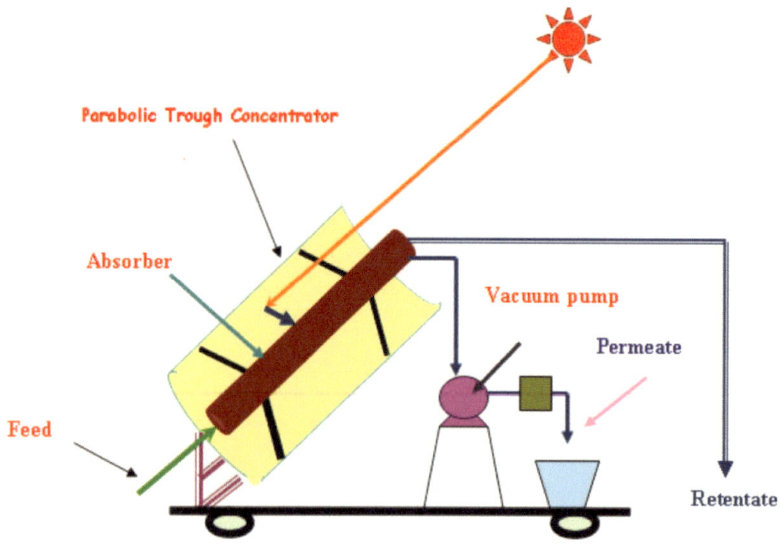

Figure 12. Solar-VMD installation.

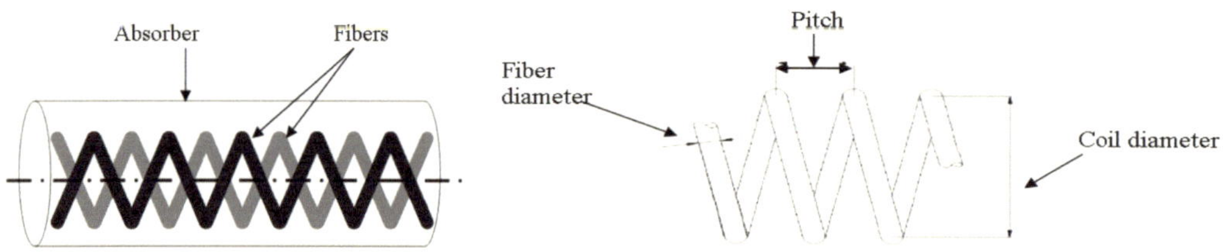

Figure 13. Basic geometry of a helical fiber.

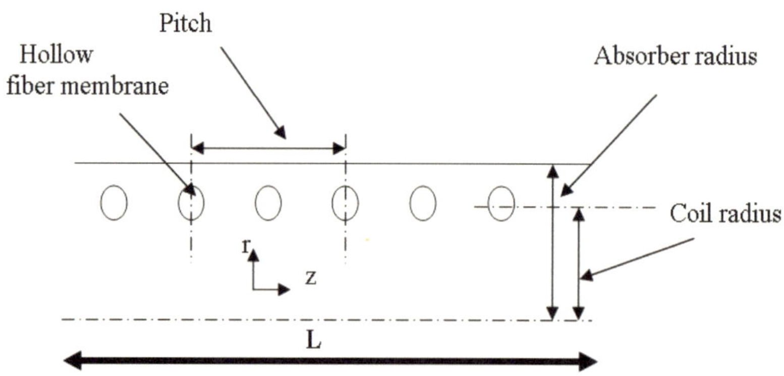

Figure 14. Domain of study [44].

3.2. Mathematical model

Based on **Figures 13** and **14**, we developed a mathematical model and the following assumptions are used for the numerical calculations:

1. The flow is fully developed before it enters the inlet of the absorber.
2. Fluid is incompressible and Newtonian.
3. The motion is considered as axisymmetric, hence, only half of the absorber is considered.
4. The gravity force is neglected.
5. All angular gradient parameters are negligible; the model is described in the coordinates r and z.
6. No slip condition is valid on the surface of the fiber.
7. All simulations are carried out assuming steady state.

Under these conditions, the appropriate governing equations are written [45]:

Continuity equation:

$$\frac{\partial u_r}{\partial r} + \frac{u_r}{r} + \frac{\partial u_z}{\partial z} = 0 \tag{1}$$

Momentum equations:

$$u_r \frac{\partial u_r}{\partial r} + u_z \frac{\partial u_r}{\partial z} = \nu \left[\frac{\partial}{\partial r} \left(\frac{1}{r} \frac{\partial r u_r}{\partial r} \right) + \frac{\partial^2 u_r}{\partial z^2} \right] \tag{2}$$

$$u_r \frac{\partial u_z}{\partial r} + u_z \frac{\partial u_z}{\partial z} = -\frac{1}{\rho} \frac{\partial P}{\partial z} + \nu \left[\frac{1}{r} \frac{\partial}{\partial r} \left(\frac{r \partial u_z}{\partial r} \right) + \frac{\partial^2 u_z}{\partial z^2} \right] \tag{3}$$

Energy equation:

$$u_r \frac{\partial T}{\partial r} + u_z \frac{\partial T}{\partial z} = \alpha \left[\frac{1}{r} \frac{\partial}{\partial r} \left(r \frac{\partial T}{\partial r} \right) + \frac{\partial^2 T}{\partial z^2} \right] \tag{4}$$

The boundary conditions for velocity and temperature are as follows:

At the inlet of absorber, $Z = 0$:

$$\begin{aligned} u_z &= 2u_0 \left(1 - \left(\tfrac{r}{R} \right)^2 \right) \\ u_r &= 0 \\ T &= T_{in} \end{aligned} \tag{5}$$

At the exit, $Z = L$:

$$u_r = \frac{\partial u_z}{\partial z} = \frac{\partial T}{\partial z} = 0 \tag{6}$$

At the hollow fiber membrane surface:

$$\begin{aligned} u_z &= 0 \\ u_r &= 0 \\ T &= T_{\text{inter}} \end{aligned} \tag{7}$$

At the absorber interior wall, $r = R$:

$$\begin{aligned} u_z &= 0 \\ u_r &= 0 \\ T &= T_w \end{aligned} \tag{8}$$

The temperature at the feed/membrane interface (T_{inter}) is related to the bulk temperature (T_b) by the following heat balance equation [46]:

$$J_v L_v = h_f (T_b - T_{\text{inter}}) \tag{9}$$

The latent heat of vaporization of water (L_v) is given by [47]

$$L_v = 2538.2 - 2.91\ T_{\text{inter}} \tag{10}$$

The absorber interior wall temperature (T_w) is given by the below equation:

$$T_w = T_e - \frac{q_u e}{\lambda_m} \tag{11}$$

where T_e is the absorber exterior wall temperature and q_u is the useful heat flow, given by

$$q_u = q_a - q_e \tag{12}$$

The incident power of the absorber radiance, "q_a," is expressed by

$$q_a = IC_g \rho \gamma \alpha \tau \tag{13}$$

The sum of the heat losses by convection and radiation between the absorber and the surrounding, "q_e," is given by

$$q_e = \varepsilon_a \sigma (T_e{}^4 - (T_a - 11)^4) + (5.7 + 3.8 w_s)(T_e - T_a) \tag{14}$$

The thermal energy balance equation of the absorber "i" element (**Figure 15**) is

$$q_u s_i = \dot{m}_{oi} cp T_{foi} - \dot{m}_{ii} cp T_{fii} + (\dot{m}_{ii} - \dot{m}_{oi}) L_v \tag{15}$$

The dominant mechanism of mass transfer through the membrane pores at low vacuum

http://dx.doi.org/10.5772/67653

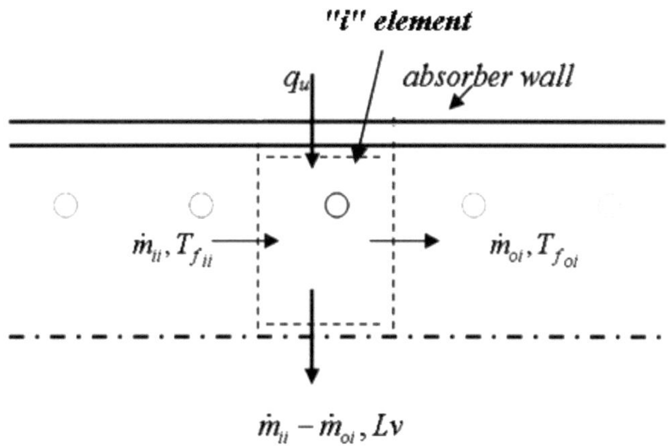

Figure 15. Thermal balance on the "*i*" element of the absorber [44].

pressures is Knudsen [48, 49]. This model suggests a linear relationship between the permeate flux (J_v) and the transmembrane water vapor pressure difference (ΔP) [10]:

$$J_V = \frac{k_m}{\sqrt{M_w}}\Delta P = \frac{k_m}{\sqrt{M_w}}(P_{\text{inter}} - P_v) \tag{16}$$

The membrane permeability coefficient (k_m) can be related to membrane structural properties such as its membrane thickness (δ), pore tortuosity (τ), and pore radius (r) [50]:

$$k_m = 1.064\frac{r\varepsilon}{\delta\tau}\sqrt{\frac{1}{RT}} \tag{17}$$

The water vapor pressure (P_{inter}) at the liquid/vapor interface may be related with the temperature, by using Antoine's equation [51]:

$$P_{\text{inter}}(T) = \exp\left(23.238 - \frac{3841}{T - 45}\right) \tag{18}$$

where P_{inter} (T) is in Pa and T is in K.

Following an analogy between the helically coiled fiber and the coiled tube heat exchanger, we used the correlation presented by Salimpour [52, 53], in order to calculate the outside heat transfer coefficient

$$Nu = 19.64\text{Re}^{0.513}\text{Pr}^{0.129}\gamma^{0.938} \tag{19}$$

$$\gamma = \frac{P}{2\pi R_c} \tag{20}$$

The helically coil length (Lc) is given by

$$Lc = \left(\frac{L+P}{P}\right)\sqrt{p^2 + (\pi Dc)^2} \tag{21}$$

For each "i" element of the absorber, we can calculate the permeate flow rate "$\dot{m}_{pi}$" according to the below equation:

$$\dot{m}_{pi} = J_{vi}\pi Lc_i d_o \tag{22}$$

The temperature polarization coefficient, TPC, is defined to measure the effective temperature difference between T_a and T_{am}. For VMD configuration, TPC is defined as

$$TPC = \frac{T_{am}}{T_a} \tag{23}$$

The solution procedure and the flow chart of the calculation were presented in Ref. [54].

3.3. Effect of the coil pitch

In **Figure 16**, it is noticed that the permeate flow rate increases with the increase of the coil pitch to reach their maximum toward the pitch value equaling 32.2 mm and decreases thereafter [44].

It should be noted that when the pitch decreases, the size of the wake also decreases and the empty space between the fibers available for bulk flow is deceased. This leads to a decrease of the feed velocity and the Reynolds number, which causes a decrease of the boundary layer heat transfer coefficient and the permeate flux. However, when the pitch decreases, the fiber exchange surface increases (**Figure 17**). These two variations of permeate flux and fiber exchange surface lead to obtain the optimum pitch, which is 32.2 mm.

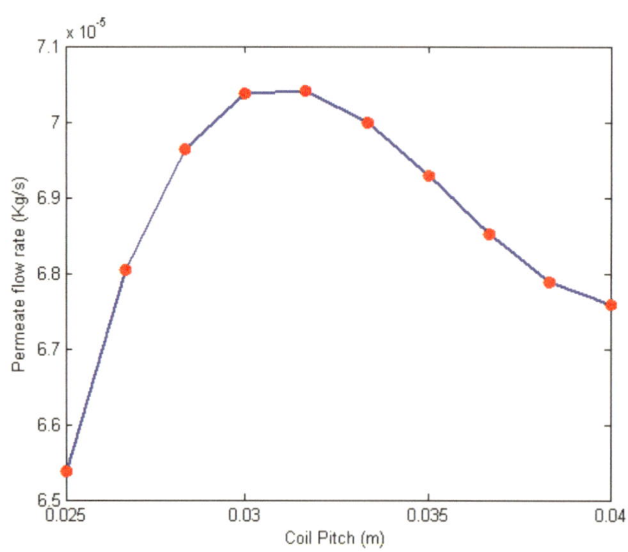

Figure 16. Effect of coil pitch on the permeate flow rate.

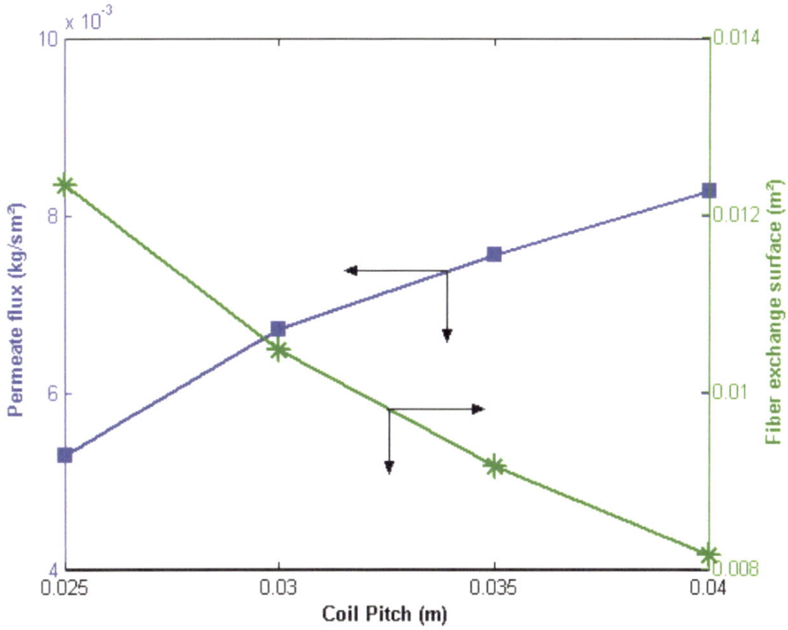

Figure 17. Effect of coil pitch on permeate flux and fiber exchange surface.

3.4. Effect of the distance between fiber and absorber internal wall

In order to optimize the distance between fiber and absorber internal wall, we have used the value of 32.2 mm for the coil pitch. After simulation, the obtained results are shown in **Figure 18**. In this figure, the reliance of the permeate flux on the coil radius is clear. As seen, the permeate flux increases sharply to reach their maximum toward the value of 4.3 mm for the distance between fiber and absorber internal wall (95.7 mm for the coil radius) and decreases thereafter. It is important to say that this decrease is due to the decrease of velocity, which is influenced by the width of the channel between the outside face of fiber and the absorber interior wall. When the value of this width decreases, the effect of channel blockage is remarked. In addition, the parabolic profile of the inlet velocity leads to obtain decreased values of velocity near the absorber interior wall [44].

3.5. Comparison between linear and helical fiber

In order to compare the coil to the linear fibers, we conserve all fiber characteristics and we change only the geometric configuration. In the case of the linear configuration, we are interested by the domain shown in **Figure 19**. Due to symmetry, the modeled domain is reduced to that presented in **Figure 20**.

3.5.1. Effect of fiber configuration in temperature polarization coefficient (TPC).

Figure 21 illustrates the obtained data for the variations of the TPC for the two fiber configurations. These fibers are used in VMD coupled with solar energy in order to increase the temperature of the feed, which flow in the shell-side.

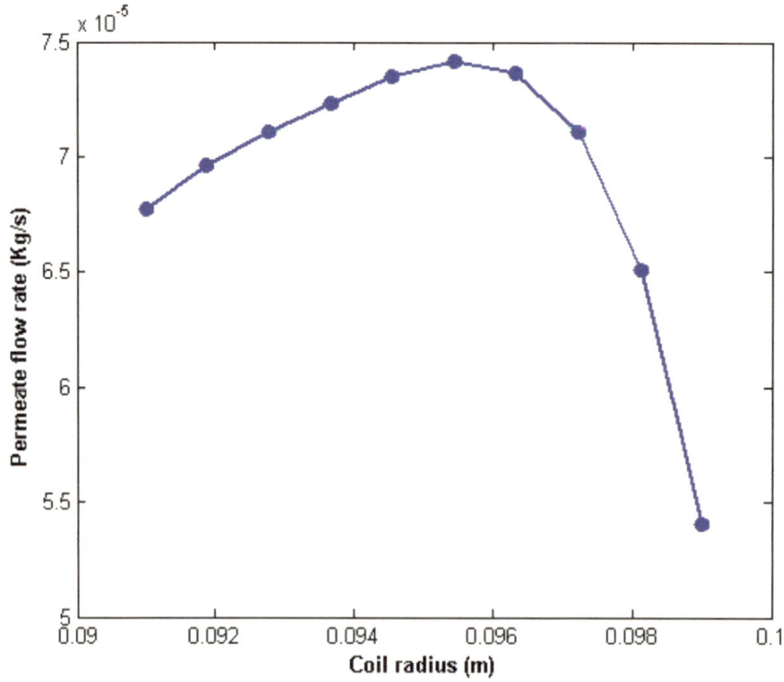

Figure 18. Evolution of permeate flow rate with coil radius.

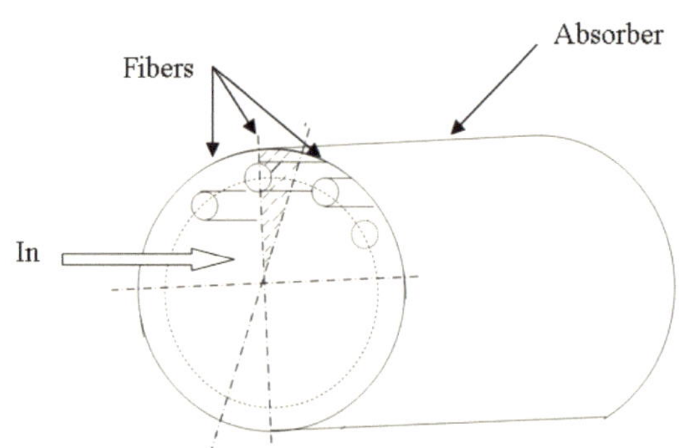

Figure 19. Linear fibers in the solar concentrator absorber.

The inlet feed velocity, in this case, was 3.4×10^{-4} m/s corresponding to a Reynolds number of 68 and the inlet feed temperature was 20°C. According to this figure, two similar evolutions, of the TPC along the module length, were shown for the two fiber configurations. The value of the TPC drops quickly as the module length increases from 0.1 to 0.3 for linear fiber and to 0.2 for helical fiber, and then decreases slowly when the module length increases to 1. For the linear fiber, the TPC decreases first from 0.8688 to 0.8543 and reaches the value of 0.8448 when the module length is 1. Also, for the helical fiber, the decrease, in the first, of the TPC was between 0.9205 and 0.8791 to reach, at the end, the value of 0.8695. However, the TPC of the helical fiber is greater than that of the linear fiber. The improvement factor of the TPC of the helical fiber is in the range of 3–6% compared to the linear one. This improvement can be explained by the fact

http://dx.doi.org/10.5772/67653

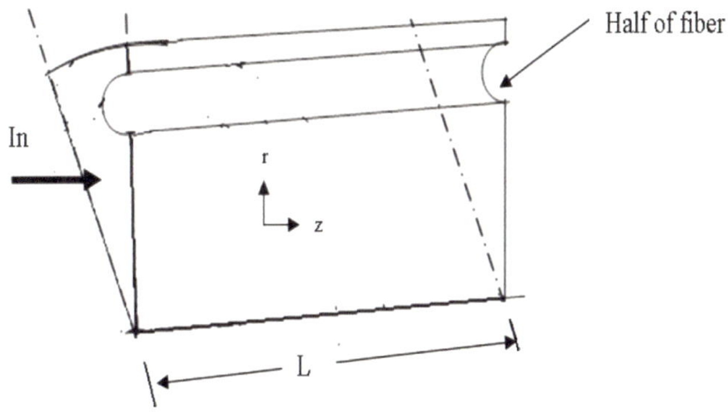

Figure 20. Domain of the study for linear fiber.

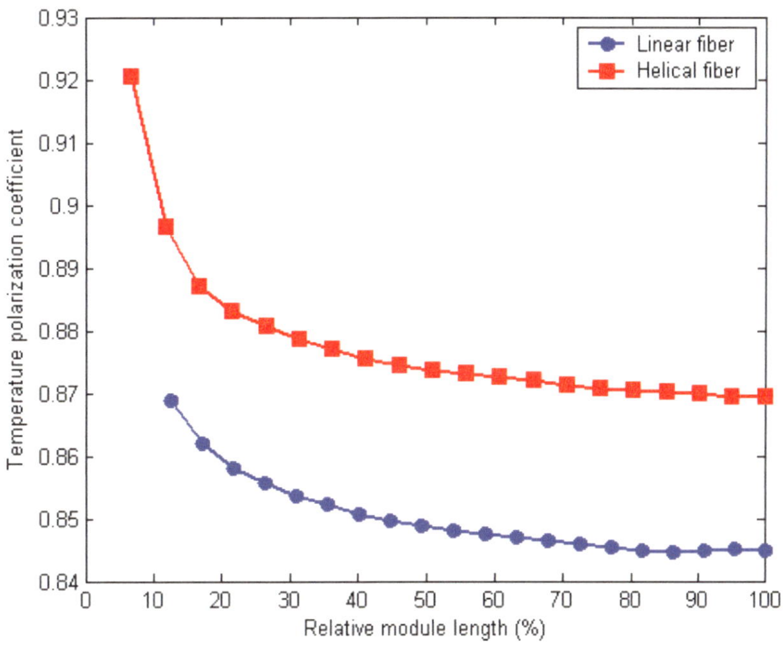

Figure 21. Evolution of temperature polarization coefficient with the relative module length for linear and helical fibers [54].

that for the flow on the shell-side of the helical fiber, a cross flow is developed, of the hot feed in the outside surface of the helical fiber, which allows to enhance the outside heat transfer coefficient. Due to this enhancement, the interface outside membrane temperature (T_{inter}), for the helical fiber, is greater to that for the linear fiber and in the same operating conditions.

3.5.2. Effect of fiber configuration in permeate flow rate

The improvement of T_{inter} leads to an increase of the permeate flow rate for the helical when compared to the linear fiber (**Figure 22**). According to this figure, the evolutions of the permeate flow rate along the module length increase from 2.7×10^{-2} to 3.58×10^{-2} kg/h for the helical

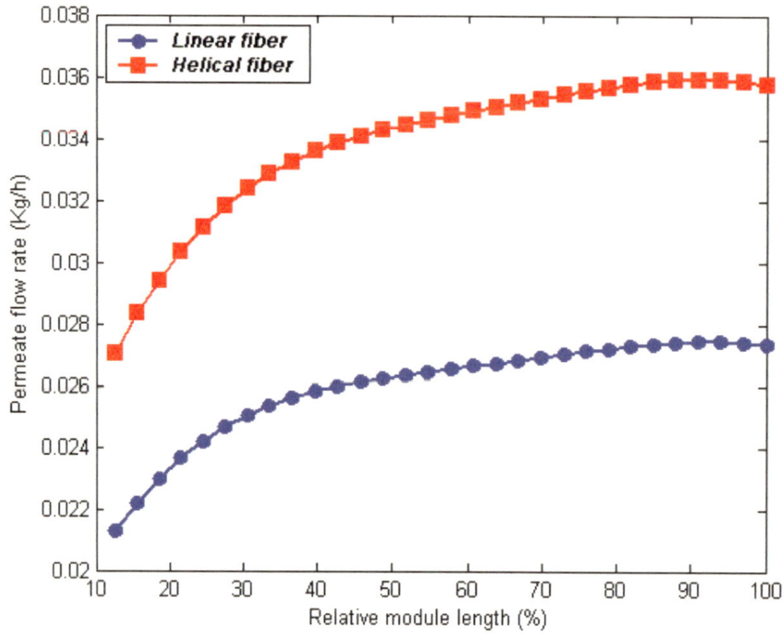

Figure 22. Permeate flow rate versus relative module length for linear and helical fibers [54].

fiber and from 2.12×10^{-2} to 2.73×10^{-2} kg/h for the linear fiber. In this case, the permeate flow rate for the helical fiber is 0.2685 and 0.21 kg/h for the linear fiber. The improvement factor in this case is about 28%. For the two configurations and along the module, the bulk temperature increases due to the solar rays focused on the exterior absorber wall. This increase in temperature raised the driving force, which is the vapor pressure difference, and the permeate flow rate. The difference between permeate flow rate for the helical and the linear fibers is principally due to the nature of the flow in the fiber outside. In the case of the helical fiber, the cross flow has an important influence on temperature polarization and permeate flow rate. However, cross flow on the shell-side yields a high heat transfer coefficient than parallel flow in the case of linear fiber.

3.5.3. *Effect of feed flow rate*

The effect of feed flow rate was investigated at the range of 20–60 l/h (Re: 34–102), while the feed temperature was maintained at 20°C. **Figure 23** illustrates the variation of the ratio between helical and linear permeate flow rates with the feed flow rate. It was found that this ratio increases strongly from 20 to 40 l/h and then increases slowly from 1.28 to 1.31 when the feed flow increases from 40 to 60 l/h. It is important to remark that for the solar membrane distillation processes, the amount of energy collected is almost unchangeable for a specific day. For this reason and for our installation, the incident solar radiation is about 800 W/m^2, when the feed flow exceeds 40 l/h, the residence time of the feed in the module decreases and the difference of the bulk temperature between the inlet temperature and the outlet temperature becomes smaller. So optimization of the feed flow rate is an effective way to obtain a high permeate flow rate in VMD coupled with solar.

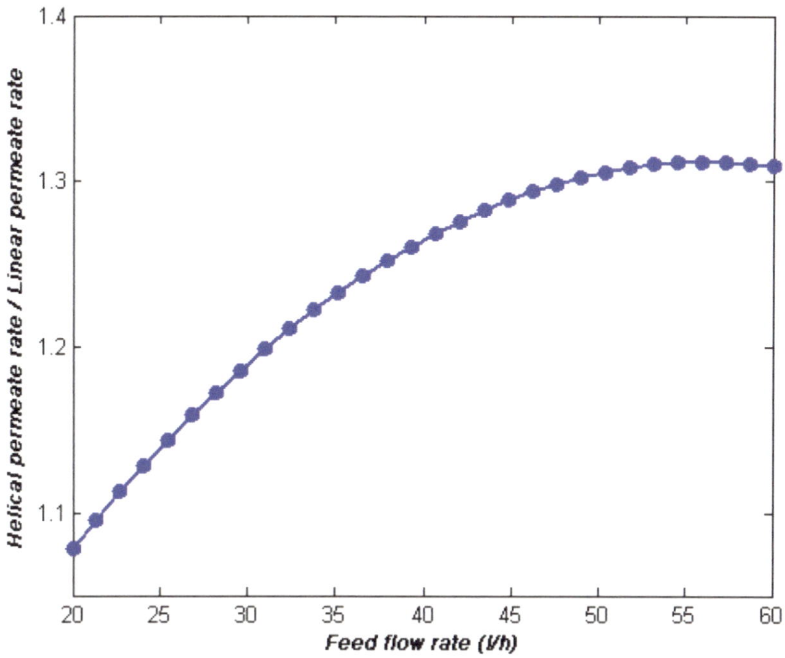

Figure 23. Effects of feed flow rate on the ratio of the helical and linear permeate flow [54].

3.5.4. *Effect of inlet feed temperature*

To obtain information about the effect of inlet feed temperature on the permeate flow rate, in both fiber configurations, feed temperature was varied in the range of 20–80°C (**Figure 24**) and the feed flow rate was fixed at 40 l/h. Permeate flow rates for both configurations showed an exponential relationship with inlet feed temperature. Although for a given flow rate, feed temperature has a small effect on the Reynolds number. There are only limited changes in viscosity and density. But the enhancement of the permeate flow rate with the inlet feed temperature can be explained by the increase in vapor pressure (Eq. (18)), or driving force (P_{inter}-P_v), with temperature. The helical fiber had a higher permeated flow rate than that of the linear fiber for all temperatures across the entire temperature range. Since the polarization coefficient temperature of the helical fiber was greater than in the linear fiber, T_{inter} in this case becomes close to the feed bulk temperature (T_b). This leads to the evolution of the permeate flow rate between 0.746×10^{-4} and 5.139×10^{-4} kg/s when the inlet feed temperature increases from 20 to 80°C.

3.6. Effect of inlet feed concentration

For our solar vacuum membrane distillation installation, when the inlet feed concentration was varied, the feed temperature and solute concentration rise along the module membrane and influence consequently the permeate flux, J_v, through the membrane. Due to this mass transfer, a temperature and solute concentration gradients are generated in the membrane liquid boundary layer. So, the bulk concentration, C_a, is different from the value at the surface of the membrane, C_{am}.

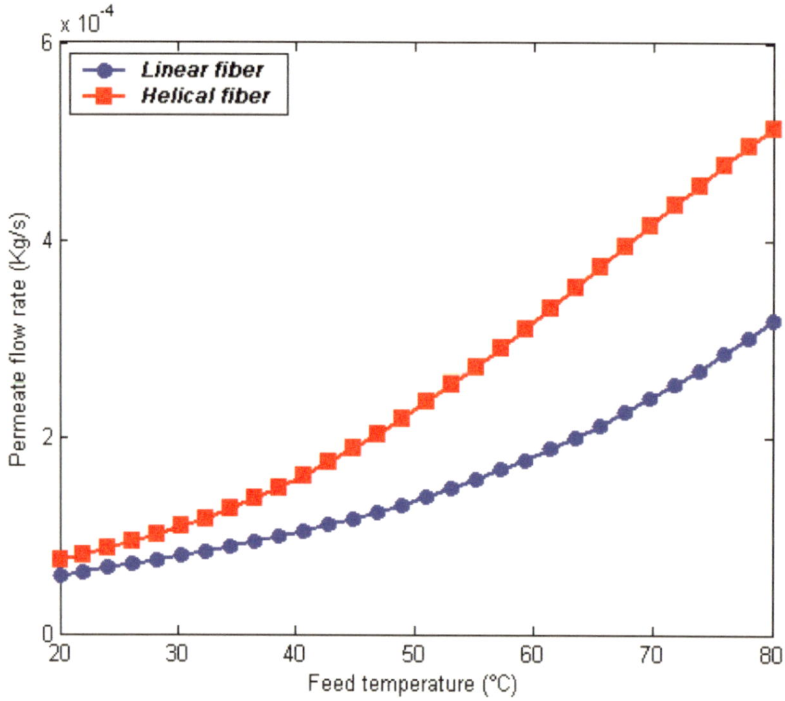

Figure 24. Influence of feed temperature on the permeate flow rate [54].

This concentration can be calculated using the film model and the mass balance across the feed boundary layer as

$$C_{am} = C_a \exp\left(\frac{J_v}{k_c \rho_a}\right) \tag{24}$$

where k_c is the film mass transfer coefficient and ρ_a is the bulk solution density. C_a and C_{am} can be used to calculate the concentration polarization coefficient, CPC, which measure the increase of solute concentration on the membrane surface

$$CPC = \frac{C_{am}}{C_a} \tag{25}$$

Figure 25 shows the effect of the inlet concentration of the feed on the permeate flux. It can be seen that the inlet concentration has a relatively small effect: increasing this concentration from 10 to 300 g/l reduces the permeate flux by only 12%. This moderate effect of the inlet feed concentration on the permeate flux is an advantage for our conception and generally for membrane distillation when it is compared to reverse osmosis system. A drop about 50% is obtained, for this process, when the inlet concentration was varied between 35 and 350 g/l [25].

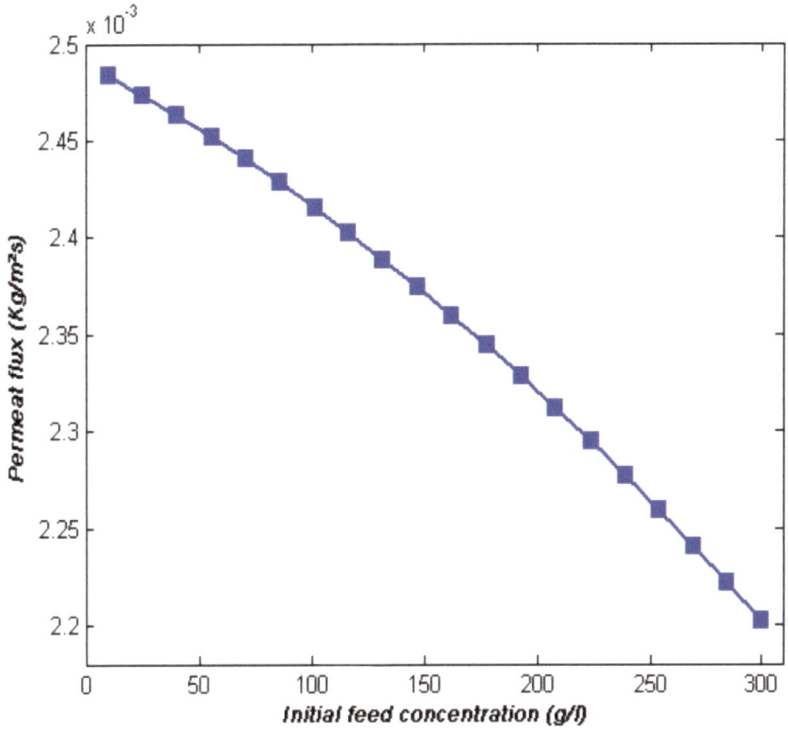

Figure 25. Effect of initial feed salt concentration on the permeate flux [55].

In order to explain the drop of the permeate flux with the increase of the inlet concentration, several phenomena can be advanced. Among these phenomena, we find the temperature polarization, the concentration polarization, and the activity coefficients.

3.6.1. *Temperature polarization*

The polarization of temperature limits heat transfer into the liquid phase. According to **Figure 26**, the temperature polarization coefficient (TPC) drops from 96.5 to 93.5% when the relative length, of the fiber, rises from 0 to 45%. Beyond this value (45%), the TPC has stabilized around 93.5%. Also, when we have varied the inlet concentration of salt between 10 and 300 g/l, no effect on the TPC was observed.

In this case, the heat transfer was not influenced by the variation of the salt inlet concentration despite the increase in temperature along the membrane module (**Figure 27**) and "T_{am}" rises with the increase of "T_a."

3.6.2. *Concentration polarization*

The concentration polarization limits the mass transfer in the liquid film. In order to investigate the influence of the inlet concentration of salt on the concentration polarization coefficient

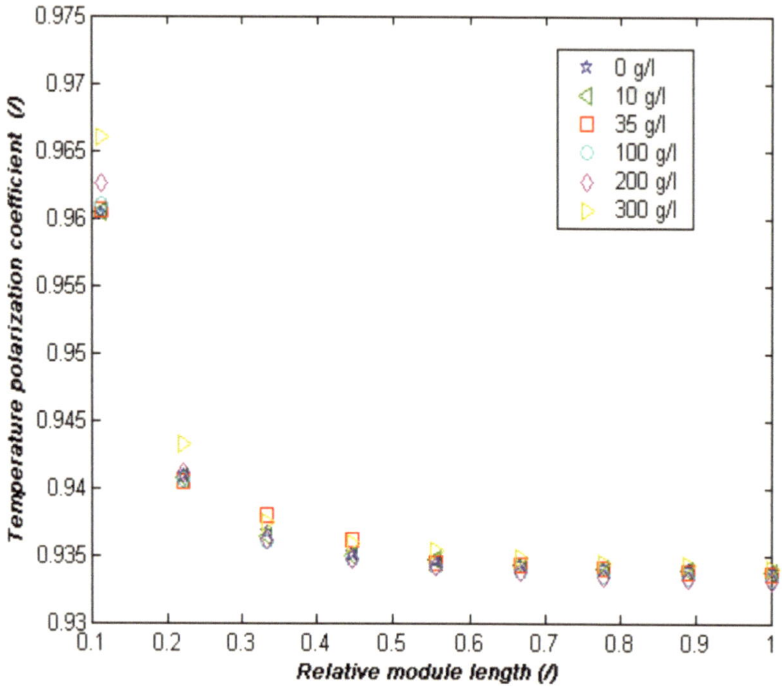

Figure 26. Temperature polarization coefficient as a function of relative module length [55].

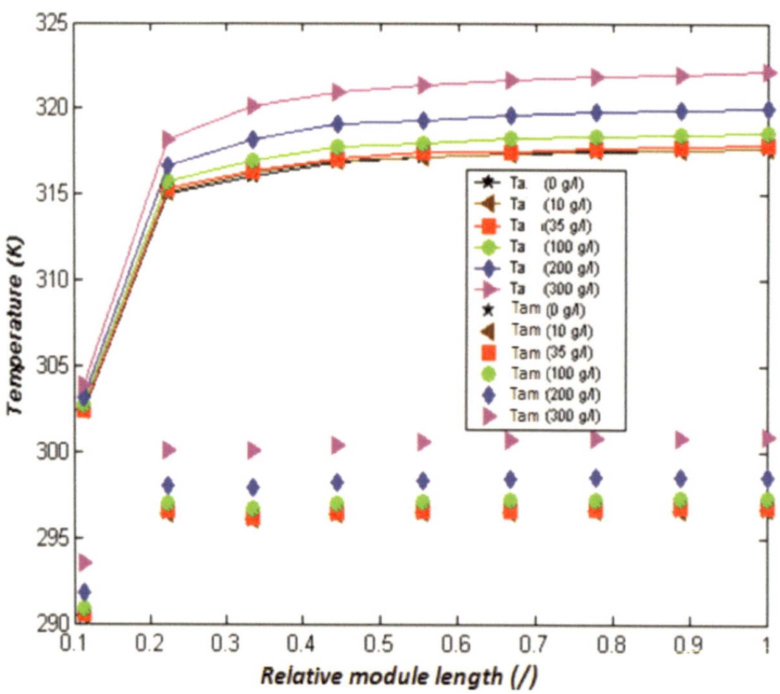

Figure 27. Bulk temperature (T_a) and membrane surface temperature (T_{am}) versus relative module length [55].

(CPC), we have plotted in **Figure 28** the variation of the CPC with the inlet concentration and along the module length.

As seen in this figure, a very small decrease of the CPC with the increase of the inlet concentration of salt is observed. This decrease is due to the decrease of the permeate flux, with the increase of the salt inlet concentration, which leads to the decrease of "$C_{am.}$"

3.6.3. Water activity

The effect of inlet salt concentration on water activity coefficient was examined by varying the inlet salt concentration between 10 and 300g/l. The obtained data shown in **Figure 29** indicate that activity coefficient is almost the same along the module length. This is because of the small value of permeate flow. However, when we varied the inlet concentration, a decrease of 20% for the activity coefficient was remarked. As a consequence, the partial vapor pressure and the driving force of the MD process decrease, which lead to the decrease of the permeate flux (**Figure 30**).

3.7. Effect of solar radiation

In order to explain the relation between the permeate flow rate and the direct solar flux, we have plotted in **Figure 31** the evolution of the permeate flow rate with the direct solar flux. According to this figure, we remarked that the permeate flow rate increases with the increase of the direct solar flux. When the direct solar flux increases, an increasing of the feed temperature is consequently reported.

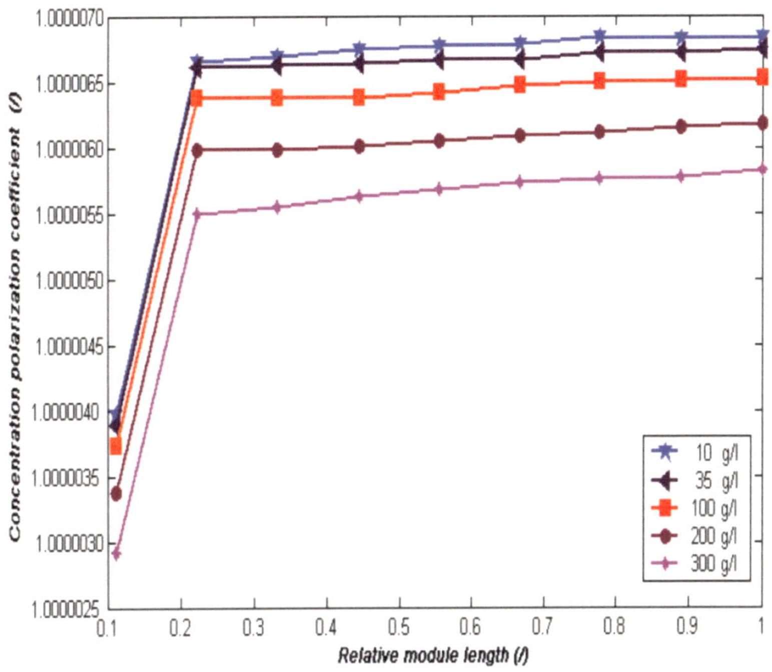

Figure 28. Concentration polarization coefficient versus relative module length [55].

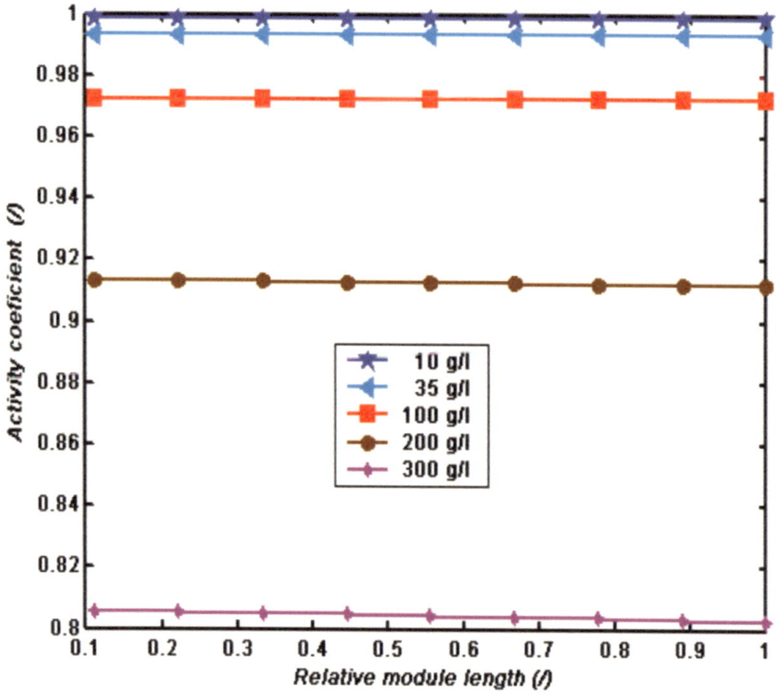

Figure 29. Activity coefficient as a function of relative module length [55].

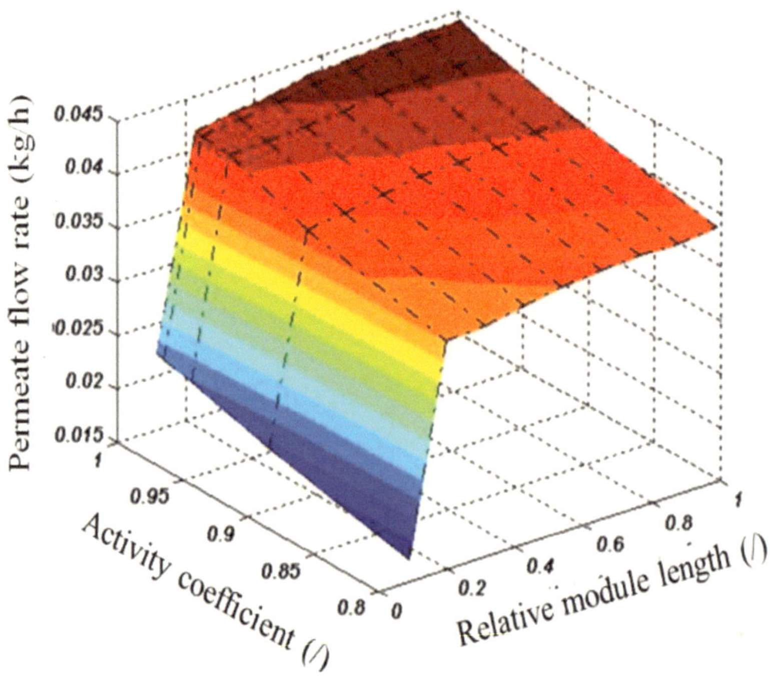

Figure 30. Permeate flow rate as a function of activity coefficient and relative module length [55].

http://dx.doi.org/10.5772/67653

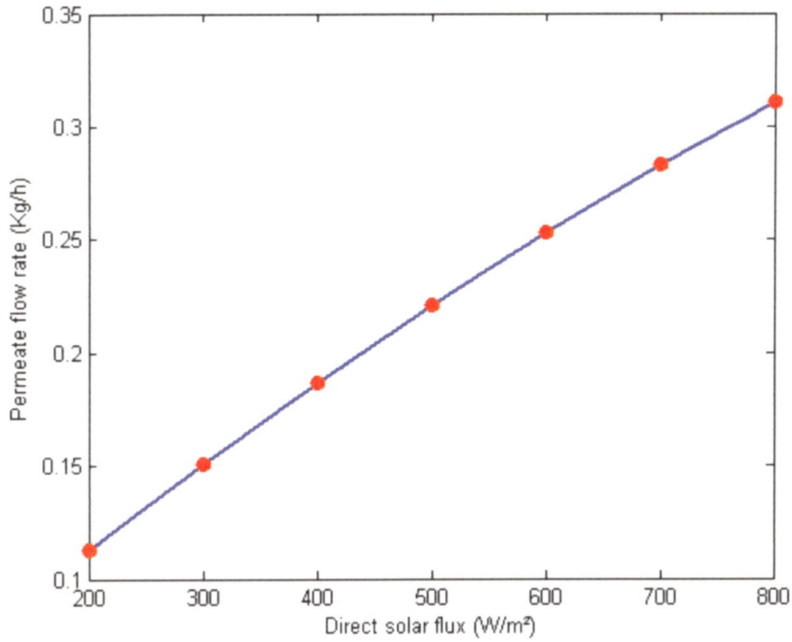

Figure 31. Effect of direct solar flux on the permeate flow rate [56].

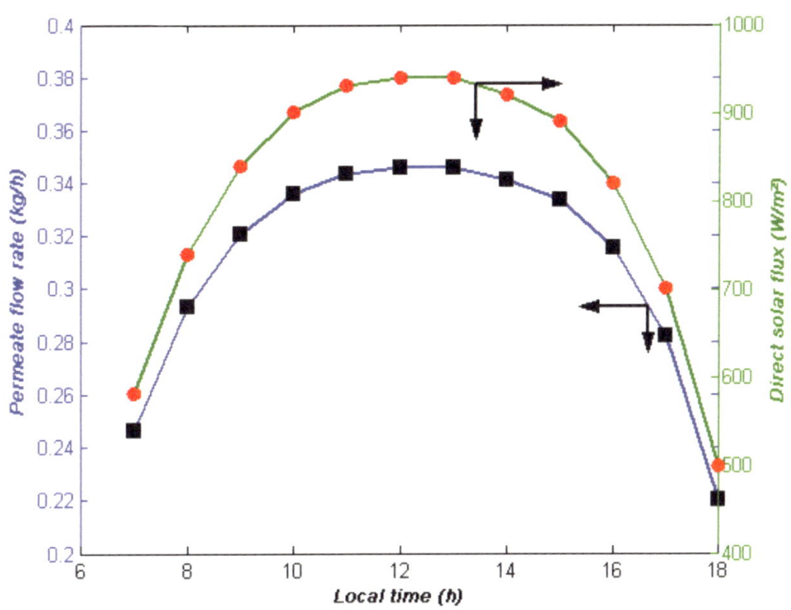

Figure 32. Permeate flow rate and direct solar flux obtained during June 21 [56].

The increase of feed temperature has a small effect on the Reynolds number at a given feed flow rate. The small change is only limited in the density and viscosity of feed. Although the increase of temperature enhances the Reynolds number somewhat, it enhances exponentially the permeate flux. This effect can be attributed to the higher water vapor sensitivity at high temperatures. This causes the increase of vapor pressure difference or driving force $(P_{H_2O} - Pv)$.

Figure 32 shows the evolution of permeate flow rate and the direct solar flux during June 21, the permeate flow rate increases in the beginning of the day to reach their maximum toward 12 h and decreases thereafter. This evolution is closely linked to the direct solar flux, which is responsible for this production and therefore has a similar evolution.

In order to determine the effect of the initial feed concentration on the permeate flow rate during June 21, we have varied the inlet concentration of salt between 10 and 300 g/l and we have plotted in **Figure 33** the evolutions of the permeate flow rate. As seen in this figure, a decrease of the permeate flow rate with the increase of the inlet feed concentration was remarked.

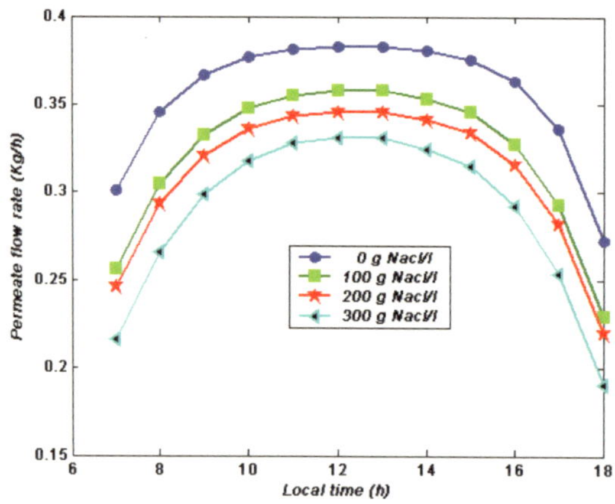

Figure 33. Effect of initial feed salt concentration on the permeate flux during June 21 [56].

4. Conclusions

Distillation process has been in continuous development during the previous decades in order to improve the process efficiency. Therefore, and in order to overcome the disadvantages of the conventional processes, the MD technique has been introduced. This technique has gained much interest, principally, for its lower energy demand and higher rejection factors. In order to improve the MD process performance, many configurations of the hollow fiber have been used. Also and according to some studies, the system efficiencies, of the MD technique, can be improved and its capital cost can be reduced when it is coupled with renewable energy. In this context, we have studied in this chapter the use of the helically coiled fiber in a solar vacuum membrane distillation installation. A mathematical model has been developed in order to describe the evolutions of permeate flow rate with the variation of inlet feed temperature, inlet feed concentration, direct solar flux, and so on. After simulation and comparison between the use of linear fiber and helical-coiled fiber, the results show an enhancement of the temperature polarization coefficient about 6% for the case of the helical fiber. This enhancement leads to obtain an improvement factor by 28% for the helical fiber and to confirm their use for membrane distillation. For the case of the effect of solar energy on the permeate flow rate, an increase of 264% is remarked for the variation of the direct solar flux from 200 to 800 W/m^2. However, a reduction of 12% for the permeate flux is obtained, when the inlet feed concentration grows from 10 to 300 g/l.

Author details

Adel Zrelli

Address all correspondence to: adel.zrelli@yahoo.fr

1 Environmental, Catalysis and Process Analysis Research Unity, National Engineering School of Gabes, Gabes, Tunisia

2 High Institute of Applied Sciences and Technology of Gabes, University of Gabes, Gabes, Tunisia

References

[1] Lazarova V, Choo K-H, Corne P. Water - energy interactions in water reuse. London: IWA Publishing; 2012. ISBN 13: 9781843395416

[2] Elbeih S F. An overview of integrated remote sensing and GIS for groundwater mapping in Egypt. Ain Shams Engineering Journal. 2015; 6: 1–15. http://dx.doi.org/10.1016/j.asej.2014.08.008

[3] Kummu M, Ward P J, Moel H, Varis O: Is physical water scarcity a new phenomenon? Global assessment of water shortage over the last two millennia. Environmental Research Letters. 2010; 5: 1-10. doi:10.1088/1748-9326/5/3/034006

[4] Pedro-Monzonís M, Solera A, Ferrer J, Estrela T, Paredes-Arquiola J: A Review of Water Scarcity and Drought Indexes in Water Resources Planning and Management. Journal of Hydrology. 2015; 527: 482–493, doi:10.1016/j.jhydrol.2015.05.003

[5] Davis J, O'Grady A P, Dale A, Arthington A, Gell P, Driver P D, Bond N, Casanova M, Finlayson M, Watts R J, Capon S, Nagelkerken I, Tingley R, Fry B, Page T J, Spech A. When trends intersect: the challenge of protecting freshwater ecosystems under multiple land use and hydrological intensification scenarios. Science of the Total Environment. 2015; 534: 65–78. doi:10.1016/j.scitotenv.2015.03.127

[6] Al-Karaghouli A, Kazmerski L L. Renewable Energy Opportunities in Water Desalination. In: Schorr M, editor. Desalination, Trends and Technologies. Rijeka, InTech; 2011. p. 149-184. DOI: 10.5772/14779.ch8

[7] Palenzuela P, Hassan A S, Zaragoza G, Alarcón-Padilla D A. Steady state model for multi-effect distillation case study: Plataforma Solar de Almería MED pilot plant. Desalination. 2014; 337: 31–42. http://dx.doi.org/10.1016/j.desal.2013.12.029

[8] Pelin O. Membrane Distillation: Principle, Advances, Limitations and Future Prospects in Food Industry. In: Zereshki S, editor. Distillation - Advances from Modeling to Applications. Rijeka, InTech; 2012. p. 233-266. DOI: 10.5772/37625. Ch11

[9] Saffarini R B, Summers E K, Arafat H A, Lienhard J H. Economic evaluation of stand-alone solar powered membrane distillation systems. Desalination. 2012; 299: 55–62. doi:10.1016/j.desal.2012.05.017

[10] Cabassud C, Wirth D. Membrane distillation for water desalination: how to chose an appropriate membrane? Desalination. 2003; 157: 307–314. doi:10.1016/S0011-9164(03)00410-7

[11] Khayet M, Matsuura T. Membrane distillation principles and applications. UK: Elsevier; 2011. ISBN: 978-0-444-53126-1

[12] Bataineh K M. Multi-effect desalination plant combined with thermal compressor driven by steam generated by solar energy. Desalination. 2016; 385: 39–52. doi:10.1016/j.desal.2016.02.011

[13] Sharaf M A, Nafey A S, García-Rodríguez L. Exergy and thermo-economic analyses of a combined solar organic cycle with multi effect distillation (MED) desalination process. Desalination. 2011; 272: 135–147. doi:10.1016/j.desal.2011.01.006

[14] Al-Karaghouli A, Renne D, Lawrence L. Kazmerski L L. Solar and wind opportunities for water desalination in the Arab regions. Renewable and Sustainable Energy Reviews. 2009; 13: 2397–2407. doi:10.1016/j.rser.2008.05.007

[15] Carayannis E G. Creating a sustainable social ecology using technology-driven solutions, 1st ed. USA: IGI Global; 2013. ISBN-13: 978-1466636132

[16] Bruggen B V, Vandecasteele C. Distillation vs. membrane filtration: overview of process evolutions in seawater desalination evolutions in seawater desalination. Desalination. 2002; 143: 207–218. doi:10.1016/S0011-9164(02)00259-X

[17] Gude V G. Geothermal source potential for water desalination – current status and future perspective. Renewable and Sustainable Energy Reviews. 2016; 57: 1038–1065. doi:10.1016/j.rser.2015.12.186

[18] Charcosset C. A review of membrane processes and renewable energies for desalination. Desalination. 2009; 245: 214–231. doi:10.1016/j.desal.2008.06.020

[19] AlMadani H M N. Water desalination by solar powered electrodialysis process. Renewable Energy. 2003; 28: 1915–1924. doi:10.1016/S0960-1481(03)00014-4

[20] Abdel-Aal E A, Farid M E, Hassan F S M, Adila E M. Desalination of Red Sea water using both electrodialysis and reverse osmosis as complementary methods. Egyptian Journal of Petroleum. 2015; 24: 71–75. http://dx.doi.org/10.1016/j.ejpe.2015.02.007

[21] Abdallaha S, Abu-Hilala M, Mohsen M S. Performance of a photovoltaic powered reverse osmosis, system under local climatic conditions. Desalination. 2005; 183: 95–104. doi:10.1016/j.desal.2005.03.030

[22] Eshoul N M, Agnew B, Al-Weshahi M A, Atab M S. Exergy analysis of a two-pass reverse osmosis (RO) desalination unit with and without an energy recovery turbine (ERT) and pressure exchanger (PX). Energies. 2015; 8: 6910–6925. doi:10.3390/en8076910

[23] Ali S M, Chakraborty A. Adsorption assisted double stage cooling and desalination employing silica gel + water and AQSOA-Z02 + water systems. Energy Conversion and Management. 2016; 117: 193–205. http://dx.doi.org/10.1016/j.enconman.2016.03.007

[24] Khawaji A D, Kutubkhanah I K, Wie J-M. Advances in seawater desalination technologies. Desalination. 2008; 221: 47–69. doi:10.1016/j.desal.2007.01.067

[25] Al-Obaidania S, Curcio E, Macedonio F, Di Profio G, Al-Hinai H, Drioli E. Potential of membrane distillation in seawater desalination: thermal efficiency sensitivity study and cost estimation. Journal of Membrane Science. 2008; 323: 85–98. doi:10.1016/j.memsci.2008.06.006

[26] Fard A K, Rhadfi T, Khraisheh M, Atieh M A, Khraisheh M, Hilal N. Reducing flux decline and fouling of direct contact membrane distillation by utilizing thermal brine from MSF desalination plant. Desalination. 2016; 379: 172–181. http://dx.doi.org/10.1016/j.desal.2015.11.004

[27] Boo C, Lee J, Elimelech M. Engineering surface energy and nanostructure of microporous films for expanded membrane distillation applications. Environmental Science & Technology. 2016. doi:10.1021/acs.est.6b02316

[28] Qtaishat M R, Banat F. Desalination by solar powered membrane distillation systems. Desalination. 2013; 308: 186–197. doi:10.1016/j.desal.2012.01.021

[29] Janajreh I, Suwwan D, Hashaikeh R. Theoretical and experimental study of direct contact membrane distillation. Desalination and Water Treatment. 2016; 57(33): 15660–15675. http://dx.doi.org/10.1080/19443994.2015.1131198

[30] Hassan M I, Brimmo A T, Swaminathan J, Lienhard J H, Arafat H A. A new vacuum membrane distillation system aspirator: concept modeling and optimization using an aspirator. Desalination and Water Treatment. 2016; 57(28): 12915–12928. http://dx.doi.org/10.1080/19443994.2015.1060902

[31] Cheng D, Gong W, Li N. Response surface modeling and optimization of direct contact membrane distillation for water desalination. Desalination. 2016; 394: 108–122. http://dx.doi.org/10.1016/j.desal.2016.04.029

[32] Moudjeber D E, -Aguirre A R, Ugarte-Judge D, Mahmoudi H, Zaragoza G: Solar desalination by air-gap membrane distillation: a case study from Algeria. Desalination and Water Treatment. 2016: 57; 22718-22725; http://dx.doi.org/10.1080/19443994.2016.113910

[33] Huayan C, Chunrui W, Yue J, Xuan W, Xiaolong L. Comparison of three membrane distillation configurations and seawater desalination by vacuum membrane distillation. Desalination and Water Treatment. 2011; 28: 321–327. doi:10/5004/dwt.2011.1605

[34] Simone S, Figoli A, Criscuoli A, Carnevale M C, Rosselli A, Drioli E. Preparation of hollow fiber membranes from PVDF/PVP blends and their applications in VMD. Journal of Membrane Science. 2010; 364: 291–232. doi:10.1016/j.memsci.2010.08.013

[35] Safavi M., Toraj M. High-salinity water desalination using VMD. Chemical Engineering Journal. 2009; 149: 191–195. doi:10.1016/j.cej.2008.10.021

[36] Li G-P, Zhang L-Z. Investigation of a solar energy driven and hollow fiber membrane-based humidification–dehumidification desalination system. Applied Energy. 2016; 177: 393–408. http://dx.doi.org/10.1016/j.apenergy.2016.05.113

[37] Banat F, Jwaied N. Economic evaluation of desalination by small-scale autonomous solar-powered membrane distillation units. Desalination. 2008; 220: 566–573. doi:10.1016/j.desal.0000.00.000

[38] Teoh M M, Bonyadi S, Chung T-S. Investigation of different hollow fiber module designs for flux enhancement in the membrane distillation process. Journal of Membrane Science. 2008; 311: 371–379. doi:10.1016/j.memsci.2007.12.054

[39] Yang X, Wang R, Fane A G. Novel designs for improving the performance of hollow fiber membrane distillation modules. Journal of Membrane Science. 2011; 384: 52–62. doi:10.1016/j.memsci.2011.09.007

[40] Yang X, Yu H, Wang R, Fane A G. Optimization of microstructured hollow fiber design for membrane distillation applications using CFD modelling. Journal of Membrane Science. 2012; 421–422: 258–270. http://dx.doi.org/10.1016/j.memsci.2012.07.022

[41] Qu D, Zhou T, Ma W, Peng Z, Li Z, Qin M. Comparison of hollow fiber module designs in membrane distillation process employed lumen-side and shell-side feed. Desalination and Water Treatment. 2016; 57(17): 7700–7710. http://dx.doi.org/10.1080/19443994.2015.1049561

[42] Mallubhota H, Hoffman S, Schidt M, Vente J, Belfort G. Flux enhancement during dean vortex tubular membrane nanofiltration. 10. Design, construction, and system characterization. Journal of Membrane Science. 1998; 141: 183–195. doi:10.1016/S0376-7388(98)00255-5

[43] Nagase K, Kohori F, Sakai K, Nishide H. Rearrangement of hollow fibers for enhancing oxygen transfer in an artificial gill using oxygen carrier solution. Journal of Membrane Science. 2005; 254: 207–217. doi:10.1016/j.memsci.2005.01.008

[44] Zrelli A, Chaouchi B, Gabsi S. Simulation of vacuum membrane distillation coupled with solar energy: optimization of the geometric configuration of a helically coiled fiber. Desalination and Water Treatment. 2011; 36: 41–49. doi:10.5004/dwt.2011.1831

[45] Bird R B, Stewart E W, lightfoot E N. Transport phenomena, 2nd ed. New York: John Wiley & Sons, Inc.; 2002.

[46] Mericq J P, Laborie S, Cabassud C. Vacuum membrane distillation of seawater reverse osmosis brines. Water Research. 2010; 44: 5260–5273. doi:10.1016/j.watres.2010.06.052

[47] Chaouchi B, Zrelli A, Gabsi S. Desalination of brackish water by means of a parabolic solar concentrator. Desalination. 2007; 217: 118–126. doi:10.1016/j.desal.2007.02.009

[48] Wirth D, Cabassud C. Water desalination using membrane distillation: comparison between inside/out and outside/in permeation. Desalination. 2002; 147: 139–145. doi:10.1016/S0011-9164(02)00601-X

[49] Izquierdo M A, Jonsson G. Factors affecting flux and ethanol separation performance in vacuum membrane distillation (VMD). Journal of Membrane Science. 2003; 214: 113–130. http://dx.doi.org/10.1016/S0376-7388(02)00540-9

[50] Mengual J I, Khayet M, Godino M P: Heat and mass transfer in vacuum membrane distillation. International Journal of Heat and Mass Transfer. 2004; 47: 865–875. doi:10.1016/j.ijheatmasstransfer.2002.09.001

[51] Banat F, Jumah R, Garaibeh M. Exploitation of solar energy collected by solar stills for desalination by membrane distillation. Renewable Energy. 2002; 25: 293–305. http://dx.doi.org/10.1016/S0960-1481(01)00058-1

[52] Salimpour M R. Heat transfer characteristics of a temperature-dependent-property fluid in shell and coiled tube heat exchangers. International Communications in Heat and Mass Transfer. 2008; 35: 1190–1195. http://dx.doi.org/10.1016/j.icheatmasstransfer.2008.07.002

[53] Salimpour M R. Heat transfer coefficients of shell and coiled tube heat exchangers. Experimental Thermal and Fluid Science. 2009; 33: 203–207. http://dx.doi.org/10.1016/j.expthermflusci.2008.07.015

[54] Zrelli A, Chaouachi B, Gabsi S. Simulation of a solar thermal membrane distillation: comparison between linear and helical fibers. Desalination and Water Treatment. 2014; 52(7–9): 1683–1692. doi:10.1080/19443994.2013.807033

[55] Zrelli A, Chaouachi B, Gabsi S. Impact of the feed concentration on the permeate flux of the solar vacuum membrane distillation equipped with helically coiled fibers. In: Proceeding of IEEE Renewable Energy Congress (IREC), 2014 5th International. doi:10.1109/IREC.2014.6827014

[56] Zrelli A, Chaouachi B, Gabsi S. Use of solar energy for desalination by membrane distillation installation equipped with helically coiled fibers. In: Proceeding of IEEE Renewable Energy Congress (IREC), 2015 6th International. doi:10.1109/IREC.2015.7110893